Urban Wilderness

Urban Wilderness
a guidebook to resourceful city living

Christopher Nyerges

Illustrated by Janice Fryling

PEACE PRESS

©1979 Christopher Nyerges
All rights reserved.

Peace Press
3828 Willat Avenue
Culver City, California 90230

Cover drawings by Catherine Kanner
Typesetting by Freedmen's Organization, Los Angeles
Printed in the United States of America by Peace Press

The illustrations on pages 105, 109, 110, 112, 116, 128, 129, 133, 143, 226, 227 first appeared in *A Southern Californian's Guide to Wild Food,* ©1978 Christopher Nyerges, White Tower Inc. Press.

The chart on page 26 is reprinted with permission from Farallones Institute.

Library of Congress Cataloging in Publication Data

Nyerges, Christopher.
    Urban wilderness.
    Includes bibliographies.
    1.   Home economics.   2.   Pollution.
3.   Organic gardening.   4.   Wild plants, Edible.
5.   Urban flora.   I.   Title.
TX147.N93   640   79-1565
ISBN 0-915238-28-4

1 2 3 4 5 6 7 8 9   85 84 83 82 81 80 79

To the uncorrupted, temporarily small-bodied persons of the world, sometimes referred to as children.

CONTENTS

Introduction	1
I. HOUSEHOLD ECOLOGY	**5**
A Nation of Wasters	6
Recycling Household Items	11
One Product with Many Uses	22
An Ecological Home	25
The Energy-Efficient Home	29
Bathroom Water Conservation	41
II. CITY GARDENING	**47**
The Container Garden	48
The Kitchen Garden	57
The Edible Ornamental Garden	62
Landscaping With Herbs	68
The Drought-Resistant Garden	80
Worms, Snails and Other "Pests"	85
III. WILD CITY PLANTS	**95**
Assault on the Weed	97
A Wild Thanksgiving	101
Gathering Urban Weeds	103
Water Plants	126
Food Grows on Trees	140
Mushrooms	149
Plant the Earth	153

IV. THE PROBLEM WITH POLLUTION — **155**
    Aluminum Poisoning — 156
    Garlic Saves — 161
    Insecticides — 165
    Mind Pollution — 171
    Bicycles — 177

V. CITY SURVIVAL — **183**
    First-Aid — 184
    Emergency Weather — 189
    Shelter — 194
    Fire — 198
    Survival Clothing — 202
    Time and Direction — 208
    Sights, Smells and Sounds of Night — 214
    Weather Forecasting — 219

ACKNOWLEDGMENTS

A book of this nature is not the work of one person. The book you hold in your hands is but the tip of the iceberg; it represents the fruits of countless pioneers and researchers throughout history, as well as those who regularly recorded this information. Many people's efforts were required to produce this book. For typing, Edward Johnson, Janice Fryling, and Barbara Elman; editing, Evelyn Spire; copy-editing and proofreading, Evelyn Spire and Dorothy Schuler; proofreading the original manuscript, Richard E. White. For her artwork and support, Janice Fryling.

I wish to thank all who have helped make this book a reality. Most of all, I wish to thank all who daily practice an ecological way of life, in both their thinking and their actions, and by so doing, help and encourage others along the way.

The book itself is of no value other than to plant the seeds within each of us which lead to positive and constructive daily actions.

INTRODUCTION

In the wake of the intensified ecological awareness of the 60s, many of the followers of this back-to-earth movement decided they could no longer live in the cities. They departed, almost en masse, to various pastoral locales to seek their peace, solitude and natural living.

All too often, the move to the country is motivated by escapism. There *are* many dangers in the city, and escape often appears to be the only option.

Soon after some acquaintances made their big move out of the urban rat race, they suffered a curious and unexpected response: guilt and boredom. Why should guilt and boredom result from the move to an idyllic setting in which they could live purer, more natural lives? Should not the result be joy, fulfillment, and serenity? Obviously, something was amiss.

The guilt results from being in the wrong place at the wrong time and knowing your work is elsewhere. The boredom is the painful consequence of an unprepared, premature move to the country.

Generally, not long after the boredom and guilt set in, the idealistic country-dazed ecologists swallow their pride and start making their way back to the urban jungle. This return is almost invariably accompanied by an acceptance (albeit unwilling) of all the former wasteful habits that seem to go hand in hand with city living.

This is not to say that those of us who have been brought up in the city can't make it in the country. It simply says that merely changing our geographic position from one of high human density to one of low human density will, in itself, do nothing to change our *attitude* or actions. In the city we have acquired an attitude that offers this rationale: "It's okay to waste. What do you expect me to do, anyway? Besides, everyone does it, and someone is paid to pick it up."

Our attitude makes all the difference in the world in the way we treat our environment, be it urban or rural. I have often heard words to this effect: "Why bother worrying about recycling, gardening, and all that stuff here in the city? Everything is so screwed up that it really won't make any difference. But once I get my mountain cabin, things will be different. I'm going to have a wood stove, grow all my own food, raise animals. Yesiree, I'm going to. . . ." Sound familiar? Of course it does. Tomorrow is always a better day to change our ways. Problem is, tomorrow never comes. This book shows us how to change, here and now, in the city.

Living lightly on the earth can add meaning and value to our lives. As city dwellers we take it for granted that someone else is going to provide for all our needs. Especially for us, becoming more self-sufficient can revolutionize our lives and our perspective on our relationship to the world. Once you begin to practice the suggestions for a more ecological city life I hope you'll move along to such essential questions as, "Who am I?" and "What am I doing on this planet?" Obviously, these are beyond the scope of this book, but are essential areas of exploration.

I take no stand either for or against city or country living. They both have their highlights and their pitfalls. Either may be good or bad, depending on various circumstances and your particular point of view at any given time. I write with the city dweller in mind because most of us live in the cities today. Most of us breathe the smog, fight the rush hour traffic, lock all our doors and windows, have televisions to entertain us, and are pacified by the garbage disposal, no-pest strips, and deodorants.

Scores of books will tell you how to make it in the country. Here's a book on how to make it in the city. I'm not encouraging (or discouraging, for that matter) people to live in the city. I'm merely showing that *if*, for whatever reason, you now live in the city, here is a way to do it more ecologically.

This book exposes the lie that we can't live in the city and live ecologically at the same time. If you wish to have personal proof, use the guidelines within these pages and prove it by doing.

Some of the solutions proposed are extremely simple ideas. Don't let their simplicity fool you. A wise man once said, "Take care of all the little problems and the big ones will take care of themselves." But don't believe me — try for yourself.

Whether you are in the city by choice or otherwise, here is a plan of action to change your attitude. And take note: even the country farmer will find much value here. This book has something for everyone.

You'll also notice that I cover wilderness survival. It's not out of place at all. In fact, many wilderness survival instructors have begun openly stating that they are teaching their students how to survive, "in the event of urban breakdown or catastrophe." This is an important point. Various disasters that can and do occur regularly and obliterate the fine line between wilderness and city survival. Some day our body limits may be tested and our survival will depend on our personal preparedness. We need the lessons of the woods as a key to survival—survival in the urban wilderness.

For those of us whose home is the city, we must realize that it is an error in thinking to believe that we *must* depart to the less-populated areas before we can live ecologically. We can begin here and now. It all begins with the decision to do so. A change in our lifestyle can come with little or no expense. The change is primarily one of attitude. We need to abandon the apathetic, disinterested posture of "I can't change anything—I'm just one person."

Country living can be experienced here and now. No, it doesn't require an exceptionally large piece of land. But it does require that self-sufficiency and full use of all our resources be valued as much as —or more than—convenience.

Readers: I welcome all your questions, suggestions, and personal experiences and comments. May this book guide you in your exploration of a better life in the Urban Wilderness and may it allow all your camels to reach Mecca.

I
HOUSEHOLD ECOLOGY

We are certainly capable of using our energy and resources efficiently, conserving what is available to us, and conducting our lives more ecologically. Recycling household items saves money and resources, and gives us control over our refuse.

Electrical appliances should be used thoughtfully and sparingly. A full washload saves kilowatt hours of electricity; manual appliances, such as can openers, are preferable to unnecessary electrical appliances. Learning to cook with the sun's energy, rather than the gas or electric company's, prepares us with alternative survival capabilities. It is knowledge we can use anywhere, as long as the sun continues to shine. The sun's energy is renewable and free.

Human-powered tools such as the treadle sewing machine, have been around for a long time. These tools provide free and unlimited energy, and beneficial exercise for the operator. The EnergyCycle Workhorse, for example, will generate electricity for chores around the house, garden and workshop.

Bathroom water usage is a key conservation issue. Our bathrooms can be designed to use water much more efficiently. There are already alternatives to our present flush toilet which are being used successfully in other countries. Shower and bath water can be recycled for various uses.

A NATION OF WASTERS

When I was young, my mother would say, "Eat all the food on your plate, people in Asia are starving." Although I felt that cleaning my plate would in no way affect Asians' empty stomachs, I always finished my meals. I interpreted her statement to mean that I should be grateful to have food, when others in the world are starving.

Since then, I've seen with crystal clarity that we are not only a nation of plenty, but a nation of wasters. A most upsetting experience occurred a few years ago after I attended an outing in the local hills. My friend Drewford Devereux had shown his students how to recognize the edible wild plants that had sustained the Indians. After we dropped off the students, we looked in the school's large trash bin. What we saw shocked us.

The trashcan contained unwrapped peanut butter sandwiches, meat sandwiches, cheese and jelly sandwiches, apples, oranges, tangerines, bananas, and even some unopened packages of potato chips. We had heard of Charlie Manson and his clan scavenging through trashcans to survive, but this was different. We were freely offered an unexpected gift, and we joyously accepted. We wondered why all this perfectly good food had been thrown away.

Since that eye opening day in my naive past, I have become acquainted with the fleeting but persistent population of trashcan food collectors, because I too have begun to check the rear of supermarkets for edible food. To the younger generation those who survive on supermarket discards are either admired as near-heroes who have beat the system artfully, or abhored. To most of society they are pitiful and scorned as bums, alcoholics and scavengers.

Just who are these people whose hands reach daily for that slightly bruised tomato or the potatoes too large to sell? Are they young or old, rich or poor, male or female, employed or unemployed? The answer is all of the above, with an emphasis on the elderly with fixed incomes. I have met all types of people around trashcans.

TRASHCAN SURVIVALISTS

A late model bronze Cadillac pulls up behind the supermarket. I have already collected two boxes of old tomatoes, celery, radishes, cucumbers, potatoes, and oranges that, for one reason or another, are unfit to sell. Out of the Cadillac steps Saul, well-dressed, middle-aged and smiling. We have never met before. As we both inspect the trashcan, we jokingly discuss what's on tonight's menu. He tells me that after his wife died, he travelled over most of this "Land of Plenty." He says he has needed to purchase produce only occasionally. "Why should I," he asks, "when they throw this stuff away?" He holds up a large tomato and laughs.

More typical of the trashcan survivalists is Paula, five feet tall, soft spoken, in her late 60s or early 70s. She is a widow with a fixed income and a face of a thousand wrinkles. Her timing is perfect; she learned the store's produce throw-away schedule long ago.

I arrived first and had already gathered most of the better discards. She told me she was gathering food for her chickens, but I knew the food was for her. I told her about the delicious meal I made the night before. She knew there was no reason to be embarrassed in my presence. I gave her some onions and beets, and dug around for a good head of lettuce.

When I see Mission Joe, I collect a few produce items for myself, and then I help him fill his boxes. His face is unshaven, his clothes are dirty, and his old and dirty automobile loudly proclaims the brightly painted message: "Jesus Saves—Read the Bible." He tells me he collects food for "the mission in Pasadena."

Occasionally I take advantage of the free food in trashcans, but I have never had to depend upon this food for survival. However, a good majority of the city's garbage food collectors depend on these discards.

Most stores frown upon people freely gathering from their bins; they prefer people purchase food inside. One produce man once told me that all the old vegetables in his bin are regularly covered with a poison chemical to discourage both flies and food collectors. Through several questionings of store officials, I learned that his story was pure fabrication, designed to keep food foragers at bay. I wondered why he would be concerned, if people take advantage of food he had already deemed worthless?

I regularly conduct wild food hikes, and I have seen how wild and abundant food plants go unnoticed and unused. Now I see that so much food is being wasted daily in our cities as well. I once collected some wild edible plants and got a large bag of too-large-to-sell potatoes behind a store. My meal that night was delicious, satisfying and free. It contained both nature's surplus and society's waste.

Markets are not the only place where one can find perfectly

healthful and edible food discarded carelessly. Dr. William L. Rathji from the University of Arizona researched trashcans in the Tucson area to see how much food families threw away. He used his data to make some national estimates of family food waste. Dr. Rathji figures that American families throw away between eight and twenty percent edible food, at a cost of $4.5 billion annually (almost as much as the Federal government spends annually on food stamps and child nutrition programs).Rathji concluded that the average family wastes at least $150 per year in food. "Homeowners go out of their way to save pennies at the store, and then don't realize that waste of edible foods adds up to much more at home."

Mothers today may want to tell their children, "Eat all the food on your plate, people in the United States are starving."

RECYCLING IS THE ANSWER

During a wild food hike I was conducting, the group spotted a discarded green wine bottle. "Look at the trash," someone said scornfully. "No, it's a good rolling pin," I said, as I picked it up and pretended to roll dough with it. "Or an attractive candle holder," said the elderly man next to me, as he took it out of my hands and began to examine it. "Or even a musical instrument," said a young girl, taking the bottle and blowing over the top. "It's a weapon," someone said, grabbing the bottle by the neck. "Make it into a drinking goblet by cutting the top off," came a voice from the rear. As everyone began to visualize a green goblet on their dining table, a youngster took the bottle and looked through it, excitedly exclaiming, "It's a pirate's telescope!" "It's money at the reclamation center," someone added, with eyes registering dollar signs.

The careless hiker who discarded the wine bottle probably never dreamed he or she would stimualte so many people's creative imaginations. Looking at each item of "trash" with this same attitude, we may never need to throw "trash" away again. Can the solid waste problem of American cities have such a simple solution? Can the urban wilderness really become self-sufficient? I'd say it all depends on you and I, and not the other guy.

Recycling is an integral part of survival. Let's examine the items we refer to as "trash." Possibly "refuse" would be a better word; we refuse to find ways to reuse and recycle items in new ways when we are finished with them. Are we lazy, dumb or both?

Recycling is not a quickie solution to "all that useless stuff I throw away;" recycling is an attitude, a way of thinking. True recycling begins in the stores and markets, before you bring an item into your home, when you decide which items to purchase. Some items have less packaging than others, some have packaging which is more biodegradable and some are easier to recycle. Look carefully and think before you buy.

Recycling can be part of the daily activities at home—not a weekend "chore" of sorting through trashcans to see what can be reused. An easy way to recycle items is to sort them, at the time you are handling them, into the various categories of recycling.

Space may be a problem for apartment dwellers. "Where am I supposed to store all this stuff?" you may ask. The answer will vary, depending on your particular circumstances, but seek out ingenious uses of available space. Maybe by rearranging things you would discover that you are not presently using your space as efficiently as possible. Talk with your apartment neighbors about setting up a central recycling center, an unused garage space or large closet. It can be done!

Recycling trash from the cities by high-technology methods has proven too costly and energy-consuming for the energy returned. Hand sorting trash after it has been collected is also too costly, time-consuming and impractical. The solution, we are told periodically, is to truck the trash out to the ocean, the desert and the dump.

Each of us are karmically responsible for our own trash. Can we accept the validity of these conclusions from the sanitation department? Recycling is the answer. Successful recycling could save the deserts and canyons that are used as land fills, save the oceans and even save the cities.

A SORTING SOLUTION

If members of each household sorted their trash into the separate categories (paper, glass, metal, miscellaneous) before the trash is collected, then all trash could be recycled easily. A huge flatbed truck, divided into four sections, could replace the present-day garbage truck. If your trash is not sorted, it is not collected.

This system might take six months to a year to get into full swing with the full support of the residents. Money incentive could encourage citizens to do the sorting. The truck drivers would keep records of the households that presorted their trash. The records would be fed into computers so the rebate incentive could be deducted from the monthly bill of those who sort.

In 1976 the city of Newport Beach, California made $50,000 from waste paper alone. Numerous successful trial programs are being funded on a small scale. Now is the time to begin implementing recycling ideas on a wide scale. Throwing our valuable organic refuse into the trashcan, because we don't know what to do with it, is senseless.

SUGGESTED READING

Sun Bear, Wabun, Nimimosha and the tribe. *The Bear Tribe's Self Reliance Book.* Bear Tribe, P.O. Box 9167, Spokane, WA 99209. A practical how-to book for those wanting to live more ecologically, with poetry, philosophy and vision. Photos and illustrations.

Swatek, Paul. *The User's Guide to the Protection of the Environment.* New York: Ballantine Books, 1970.

U.S. Department of Agriculture. *Consumer All.* House Document No. 29. Washington, D.C.: Government Printing Office, 1965.

RECYCLING HOUSEHOLD ITEMS

We have been lazy in our attitudes and actions, preferring to toss things away, out of sight, letting someone else deal with our trash. It is time to take responsibility for the waste we produce; time to change—individually and collectively—trash and waste in this country. Recycling may take extra time, but if cutting waste matters, make the time.

EGGSHELLS

The eggshells you throw away every day contain approximately 93% calcium carbonate. Calcium carbonate is used as a mineral supplement for animal foods, such as dog food and poultry mashes.

Shells must be sterilized, because several diseases common to poultry can be carried on eggshells. All commercial eggs are washed before they are sold. If you have your own chickens, be sure to wash their eggs thoroughly.

Remove as much of the adhering whites as possible; otherwise you will have rotting animal products in your eggshells. Scoop out all the white after cracking an eggshell. You will be amazed to discover how much egg solid we normally leave in the shell. Keep a tray in the oven for empty eggshells. Leave them on the tray for a few days. The pilot light is enough to dry and sterilize the eggshells. Crush the shells after drying them; mash into a powder with either a potato masher or your hands. Grind with a good quality kitchen blender. Dogs are satisfied with the kitchen blender variety of powdered eggshell. Sprinkle approximately ¼ cup on each meal.

Eggshells contain sufficient potassium, phosphorous and calcium to make them a practical fertilizer. The 4% organic matter and trace elements present in eggshells make them especially suitable for soil liming. Egg-processing plants dry and grind eggshells for fertilizer. Calcium is an essential plant nutrient which plays a fundamental part in cell manufacture and growth. Most roots must have some calcium at the growing tips. Plant growth removes large quantities of calcium from the soil, and the calcium must be replenished.

Eggshells also make an excellent fertilizer for a growing lawn. First have your soil tested. Local nurseries and mail-order garden suppliers sell simple soil testers. If you live in an area of heavy rainfall and sandy soil, the PH level may test low (below 5.5). The best gardening books recommend adding lime to a growing lawn, and calcium carbonate is best.

Prepare the eggshells as described. When they are in a fine powder form, spread them on the dry ground and work them into the soil with a rake. Use 50–75 pounds per 1,000 square feet; supplement with purchased lime, if you don't have enough eggshells. Your eggshell fertilizer will produce more alkaline soil.

Snail problems can be solved with the help of recycled eggshells. Instead of powdering the shells, crush them lightly so there are plenty of rough, sharp edges. Scatter the crushed shells in a circle around tender shoots in the vegetable garden, making sure that the circle has no breaks. When snails approach the young plant, they will have to cross the sharp eggshell barrier. The discomfort caused by the shells will make the snails retreat.

NEWSPAPERS

If you subscribe to a newspaper, you know how fast the old copies can pile up. Most of us are already aware of the many uses of newspaper. The newspaper editor may feel that the paper is a shining example of literary achievement; but once read, papers are used to swat flies, line drawers, cover books, wrap fish, clean up floors, wash windows, make paper maché, clean up grease and wipe feet on a rainy day. Newspaper provides excellent insulation. Hobos stuff their clothing with newspapers to keep warm. Newspaper can become house insulation if necessary; rolled tightly into logs, it can provide emergency heating.

After you have utilized newspaper in many ways around the house, and you still have more newspaper than you can use, recycle it. Look in the Yellow Pages of your telephone book for the closest newspaper recycling center. Many people make a tidy sum of money collecting neighborhood newspapers, taking them to recycling centers on a regular basis. Considering rising paper costs, recycling helps everyone. Schools and churches often have paper drives to raise money. Give your newspapers to these good causes.

ENVELOPES

Purchasing new envelopes is often an unnecessary waste of hard-earned dollars. Consider the mountains of advertisements that enter

Recycling Household Items 13

mailboxes every day. Many contain envelopes, presumably for your order and check. More often than not, these new and useful envelopes are tossed into the wastebasket, wasting our precious trees. (You haven't forgotten where envelopes come from, have you?) A dedicated recycler and concerned conservationist will save these envelopes and use them for personal correspondence.

If you open envelopes carefully, you can reuse them easily. Put label stickers over the old addresses, add your letter and send it off. Try blacking out the former address with a marking pen, writing the new address next to the old. If some envelopes have too much writing on them, carefully unfold the flaps, and turn the envelope inside out; tape it back together, and you have a new envelope. Occasionally an envelope will rip and tear, regardless of your care in opening it. These envelopes can be used for note paper or fire starting.

After recycling my envelopes, I write "Save Our Trees!! Recycle all Envelopes!" across the front before mailing. Several recipients of my mail have begun to do the same, at least when they write to me. Responses to envelope recycling vary from "I'll do it, but what a waste of time," to "Hey, this is a great idea." That others are now recycling envelopes as a result of my actions is important to me. Recycling begins at home.

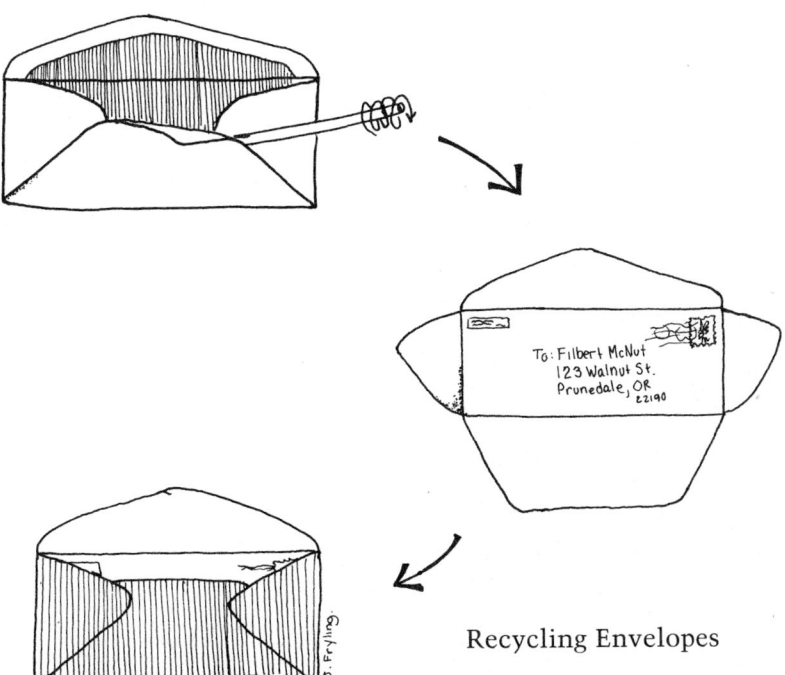

Recycling Envelopes

PLASTIC

Plastic bags can be cleaned and stored easily and have an endless array of uses. Plastic bags can usually be used many times if only we open them carefully the first time. Many manufacturers claim their plastic containers are biodegradable. Although it is not clear whether plastic causes any damage to the soil, I have seen plastic containers become brittle and break down into little bits when exposed to sunlight for a short while. When water is stored in these containers for several months, these containers actually "melt" away. I suggest you consult your yellow pages to see if there is a recycling center in your area that recycles plastic.

Plastic six-pack rings (soft drink and beer holders) are recyclable. I cannot recall having bought a six-pack of either soda or beer, but I have certainly seen plenty on the beach, in gutters, in parks and in mountain resorts; wherever people cluster and trash.

My uncle Louis in Ohio showed me a creative way of recycling plastic holders many years ago. He buys no beer or soft drinks, but he picked them off the street and saved them. When he had a sufficient amount, he attached them all together with the "twist-ems" that tie off plastic bags. He attached enough together to cover the entire outside wall of his garage. He created a unique long-lasting trellis at no expense.

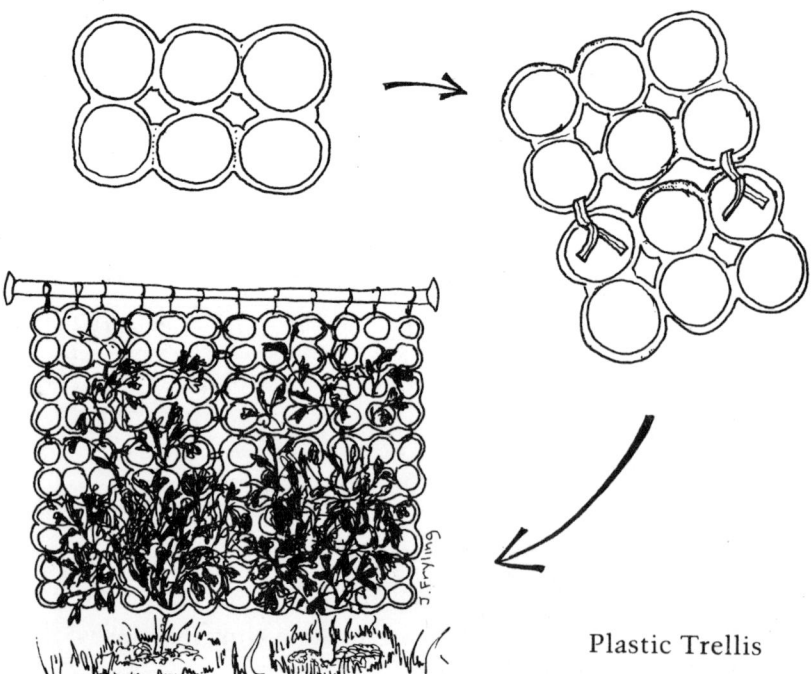

Plastic Trellis

Uncle Louis planted tomatoes along the bottom of the trellis and trained them to grow up along the wall. He used the trellis for several years with great success, creating a way to reuse an item rather than throw it away.

A plastic trellis will also support peas, beans and aromatic trailing vines. Nothing is wrong with paying hard-earned dollars for a beautiful redwood trellis. But if you want to do your part to make full use of our resources, and if a redwood trellis does not fit into your budget, then start collecting plastic six-pack holders.

BROWN BAGS

Large brown paper bags can be recycled treasures once we realize the full extent of their usefulness. Use them as trashcan liners, mailing wrappers and book covers. Uses of the versatile paper bag are endless.

Save the money you spend on drip coffee filters by making recycled paper bag coffee filters. Cut a large circle from a brown paper bag and soak it in water. Fold the circle to make a cone, and put it into the drip coffee maker.

If you don't drink coffee, you can use the paper bag drip method for your roasted grain beverages. Ground and roasted dandelion, chicory roots and barley all make a satisfying, tasty beverage, either individually or mixed.

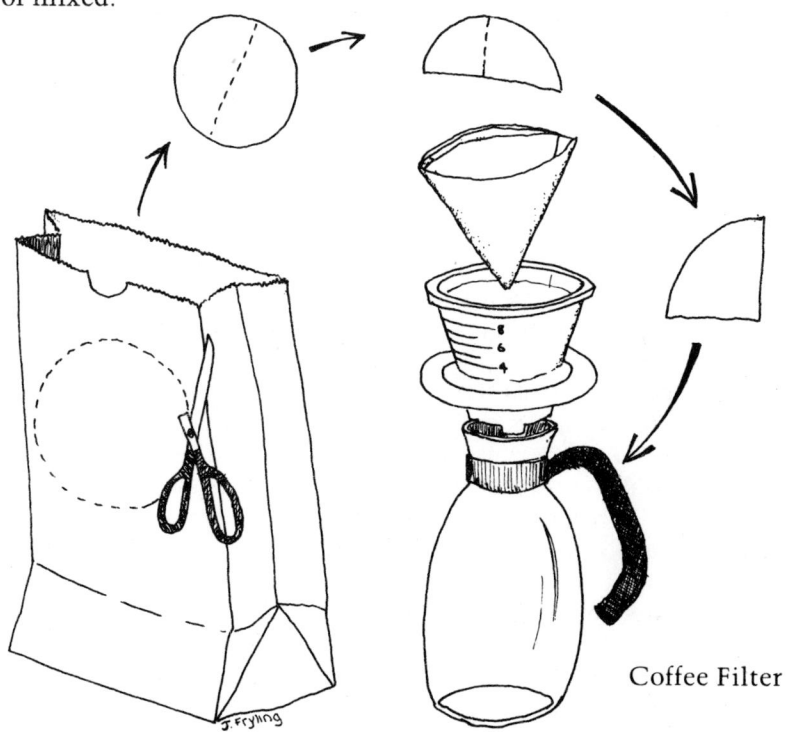

Coffee Filter

When you are finished using your filter, throw it and the contents into your worm farm compost pit. The grinds and paper will add nutrients to the soil.

Is it because we lack imagination that so many items we use are taken for granted? Consider the paper bag—a tree cut down, taken to a mill, made into paper, made into a bag and purchased by a store. All this, just so you can conveniently carry home your groceries.

SCRAP METAL

Small scrap metal pieces mixed with rain water make a unique, iron-rich plant food. Iron is a necessary nutrient for healthy soil. Curtailed chlorophyll production in plants will cause leaves to yellow and plant growth to cease. Iron is essential in chlorophyll formation. Although the actual amount of iron a plant needs is minimal, plants must have some iron.

To make iron water use only soft iron items, such as broken files, iron bolts and nuts, old hinges—anything that rusts and is no longer useful (not brass, aluminum or copper). Old steel is excellent. Put iron or steel items in an old jar (a Mason jar is ideal). Place a funnel on top to catch the rain water. Allow your brew to sit and steep indefinitely. Whenever you have rusted iron pieces, add them to the elixir. When the water level drops below the bottom third of the jar, refill with rainwater.

Using iron water is simple. Shake the bottle well, and pour a small amount of water on each plant. One cup per month will adequately feed each plant. Your plants will love their "mineral water" and smile at you in their lush, healthy condition.

TIN CANS

Although some recycling centers now take tin cans, there are a myriad of possible uses for tin cans around the house. Store nails and other small hardware in them, or use as paint containers.

A tin can will make an excellent cooking pot, dish, cup, candle holder, reflector oven, lantern or clothes washing receptacle (depending on size), for camping, emergencies and home use.

The sun will heat water in a tin can, if you paint it black. Fill a large can about 4/5 full of water, cover it, and place a large piece of aluminum foil on top. Set the can in full sun for several hours. Cover the whole works with a sheet of clear plastic, and weigh the edges down with rocks or dirt. Several hours later you will have hot water for boiled eggs, hot coffee, instant soup or washing.

Tin cans make excellent containers for melting wax. They also work as a candle mold. Tie a wick to a nail or pencil and place it in the can. Pour in melted wax and cool.

Recycling Household Items   17

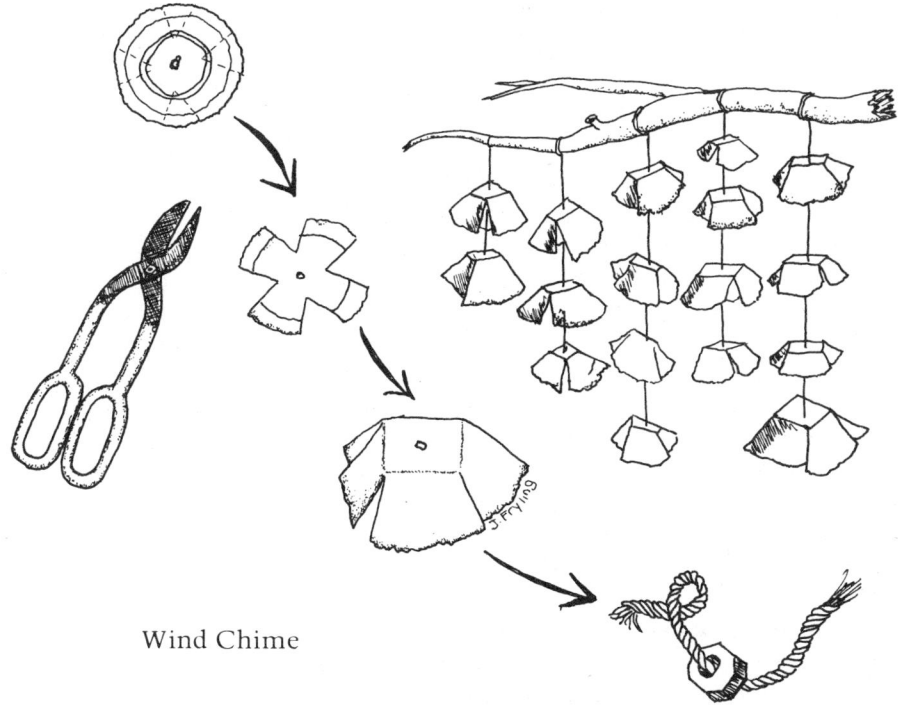

Wind Chime

Can lids can be strung together, hung up and used as wind chimes.

Gardeners often use cans as "starters" for seeds. Tin cans of all sizes make good planters. Check the container gardening section for specific planting techniques.

Youngsters can make stilts with large cans. Punch two holes toward the tops of two cans and string a sturdy cord through the holes. The cord must be long enough to use as handles. Cord length will vary with individual height.

Children love drums. Take tin cans of assorted sizes to make different sounds; turn them over and hit them with the hand or wooden spoons. Experiment with different sizes.

I have seen smokestacks for indoor woodstoves constructed entirely from old cans. One side of each can in the smokestack had several slits into which the next can was inserted. The smokestack was sturdy and had no leaks. With your imagination, and a soldering iron or epoxy glue, you can make a top-quality smokestack for your urban wilderness home.

TV TRAYS

If you purchase TV dinners in aluminum trays, there are ingenious uses for the empty trays. Tacked (or attached in some manner) to the walls of your house, they will insulate, soundproof, and block out the detrimental ultrasonic sound waves continually bombarding your home. Styrofoam or plastic egg containers can be used similarly, but they do not block out ultrasonic sound waves.

Because most urban dwellers are concerned about beauty in their homes, and many have landlords who would look dimly upon walls covered with TV trays, other solutions need to be found. Possibly one could cover the inside of the walls with old TV trays, when the house is constructed, or during apartment remodeling. You need not look at the TV trays on your walls all day, posted like some monument to the frozen food industry.

You can always recycle your aluminum cans, foils and trays. As of 1978, the aluminum industry (Aluminum Association) has established a toll-free phone number which you can call for the aluminum recycling center nearest you. The toll-free number is 1-800-223-6830. There are more than 2,000 aluminum recycling centers in the United States. The going rate for aluminum is about 16 cents per pound for aluminum beverage cans, and it takes only about 23 cans to make a pound.

I suspect that after you read the aluminum section in this book, you will seriously reconsider eating anything packaged in aluminum. Soon you may not have any aluminum to recycle. You may, however, choose to gather aluminum cans for an additional source of income. I know a young girl who gathers aluminum cans daily. She collects in the evenings after the beer-drinking crowd has gone home and left their trash for others. Gathering cans takes little time and effort, generates income and cleans up natural areas. Soon the cans accumulate, and she sells them to a recycling center. She has earned up to $30 a week recycling aluminum cans. Newspaper, cans, and bottles are equally recyclable. If we as a nation are to solve the solid waste problem, we as individuals must recognize all the uses and reuses of the things we regularly toss into the trashcan.

GLASS

Keep high-quality glass containers (such as peanut butter and pickle jars) for storing leftovers or home pickling and canning. Glass jars can help organize the workshop, because you can see exactly which container has the finishing nails, the wood screws, the washers, etc. In many workshops I have seen glass jar lids nailed or glued to a flat horizontal board, at eye level. Jars full of various hardware screw into the secured lid. This method is especially good for the cramped workspace.

For those of you with clogged toilet lines from roots growing into pipes, try pouring a cup of crushed glass into the toilet with each flush.

Glass helps cut away small roots and grime that build up inside the pipe.

Long narrow bottles make excellent rolling pins. Try putting an old light bulb inside socks while darning the holes. Be careful not to break the bulb.

Buy returnable glass containers when possible. If you cannot purchase returnable glass containers, and you have more than you can possibly use, take them to your nearest recycling center. Remember: Sort all glass before recycling. Keep three separate boxes for clear, brown and green glass in your service porch or garage.

HOUSEHOLD RECYCLING

There are many ways to recycle household items. The following is a list of creative ideas which will make recycling work for you. Expand the list yourself.

FOOD

1. Lightly stewed for dogs' daily fare (to replace kibble as much as possible).
2. Offer to other animals in descending order of intelligence/evolution.
3. Compost.
4. Spread thinly around plant or tree bases as mulch or fertilizer.
5. Under no circumstances let it stand or rot, or merely toss it onto heaps, cover it and forget it; this is the surest attractor of rats, and with rats come bubonic plague and other miseries.

PAPER

1. Fire starter.
2. Recycled constructively and imaginatively (large ice cream containers for trash baskets, etc.).
3. Mulch.
4. Turn everything possible into note paper and reusable envelopes.
5. Packing material for shipping.
6. Save for paper drives.

PLASTIC

1. Recycle in useful creative projects (targets, swimming pool floats, grain and flour scoops, planters, etc.).
2. Include in firemaking kits.

3. Recycle reusable containers.
4. Primary planters and sprouters.
5. Partially buried next to planters and trees for subsurface watering.
6. Decomposed in solar decomposition area, then crumbled and used as mulch/decompactor.

METAL

1. Recycle constructively in useful projects (wind chimes, stove flues, etc.).
2. Save thin metals for shim stock.
3. Musical instruments.
4. Separate and take "valuable" metals to junk company.
5. Primary and/or secondary planters.
6. Iron scraps to make "iron water," or mulch in iron-poor soils.
7. Storing hardware items.
8. Tin cans for small paint jobs.
9. Water reservoirs for potted plants.

GLASS

1. Recycle constructively in projects.
2. Storage containers for perishables (grains, coffee, beans, sea salt, etc.).
3. Musical instruments.
4. Break into shards and flush down drains/toilets to keep sewer lines clear.
5. Break into barrels and sell periodically to recycling centers.

SUGGESTED READING

Consumer's Marketplace Encyclopedia, American Consumer Publications, 310 E. 46th Street, New York, NY, 1979.

Egge, Ruth Stearns. *How to Make Something From Nothing.* New York: Coward-McCann, Inc., 1968. Full of ways for transforming no-longer-useful objects into something useful. Good for the home decorator.

Mother Earth News, P.O. Box 70, Hendersonville, NC 28739. $10/year.

Rich, Frank M. *Dictionary of Discards.* New York: Avenel Books, 1952.

Rosenthal, Lois. *Living Better.* Cincinnati: Writer's Digest Books, 1978. The entire book is handwritten and illustrated. A beautiful book with very practical advice for stretching your dollar in the city, recycling, wild food and consumerism. Highly recommended.

ONE PRODUCT WITH MANY USES

Baking soda is a perfect example of how one purchase at the store can apply to many home uses. Purchasing products with multiple uses in mind is the beginning of ecological thinking. Television commercials have convinced us that baking soda will deodorize your refrigerator, but were you aware of its many other uses? Learn the magic of baking soda, and apply its versatility to other products as well.

Baking soda and a little water (often mixed with salt) make one of the best and cheapest toothpastes available. (If you don't have a toothbrush, use a twig from sweet gum, dogwood, olive or any soft wood. Chew the end to soften it.) Use baking soda for bathing; it cleans well and leaves you feeling refreshed. When washing clothes, let baking soda be your natural water softener.

If you have a swimming pool, add baking soda to stabilize the water at the ideal range from 7.4 pH to 8.2 pH. Added to the pool, baking soda will help keep the eyes from burning and keep the water clear. One pound of baking soda per 10,000 gallons of water weekly is suggested.

If you've gathered wild roots for dinner and they are somewhat tough or fibrous, boil them in baking soda water and they will soften. Dump the water and reboil the roots in lightly salted water before eating them.

Baking soda makes bread rise; use it in place of baking powder. Dirty pots and pans clean more easily if you fill them with water, add baking soda and boil for a short while. Baking soda can even replace your abrasive household cleaner, for cleaning counters, sinks, bathtubs and more. It contains no bleach and will not remove the precious oils from your skin.

If your eating habits give you indigestion, heartburn or sour stomach, baking soda will come to the rescue. Put one-half teaspoon of baking soda in a glass of water and drink it down. Before reaching for baking soda (which stops the digestive process), keep the following in mind before eating: long, slow, thorough chewing is extremely important. Fast gulping is a major culprit in gas and indigestion.

If you find yourself with a badly smoking wood stove, baking soda can patch holes in the stove. Use one cup of baking soda, adding one-half cup of salt and enough water to make a thick paste. Fill the cracks and holes in the stove with the paste. The heat will harden the paste and the stove should not need further attention for some time.

A baking soda and water paste, applied to mosquito bites helps relieve itching. Whenever you have an ant problem, sprinkle baking soda along the ant route. Baking soda does not kill ants but tends to repel them. You can save money on expensive sprays and save your health as well, as ant sprays are very noxious.

Baking soda will help in the kitchen or around the campfire. Cutting off the oxygen supply, of course, is the best way to stop a fire. Cover a flaming pan of grease with a larger pan: do not use water. Although soil is your best fire extinguisher, sometimes soil is neither practical nor immediately available. Baking soda can successfully extinguish flames.

Baking soda, made into a watery paste and applied to the skin, will alleviate sunburn pain. Sore and tired feet appreciate soaking in a tub of warm baking soda water. A few tablespoons of soda in the tub is sufficient.

Baking soda has deodorizing properties and will help in many situations. Put an open box in your refrigerator at home or in the ice chest while camping. It will absorb odors. Remember to change the box every month or so. Sprinkle baking soda around outhouses to combat odors. Fill the bottom layer of your cat's litter box with one part baking soda and cover with three parts cat litter. Baking soda will absorb odor for several days and make the use of deodorizing sprays unnecessary.

Speaking of deodorants, you can make a watery baking soda mixture and wash your underarms with it. This natural deodorant is very useful on lengthy group camping trips and even at home, if you dislike spraying aluminum under your arm. Most deodorants on the market contain dangerous chemicals.

PLAY CLAY FOR CHILDREN AND ADULTS

2 cups baking soda
1 cup cornstarch
1¼ cups cold water
3 drops food coloring (optional)

Mix baking soda and cornstarch. Blend well. Add cold water and mix until smooth. Boil one minute, stirring constantly, until you have the consistency of moist mashed potatoes. Add food coloring, if desired. Spoon out onto a plate. Cover with a damp cloth and let cool.

Knead lightly and roll out on waxed paper. If you wish, you can cut out designs with a cookie cutter, bottle caps or shape by hand. Etch in a pattern. Let the art pieces dry until they are hard (24 to 48 hours). Paint with tempera or water colors, if you have used uncolored "dough." Allow the paint to dry. Coat with clear shellac or clear nail polish.

Mount pins and clips to your sculpture with white glue. Make Christmas decorations. Make animals or reliefs. Experiment on your own; make whatever you can create.

Arm and Hammer will send you free pamphlets on the uses of baking soda. Write to Consumer Relations, Church & Dwight Company, Inc., Two Pennsylvania Plaza, New York, N Y 10001.

AN ECOLOGICAL HOME

Examples of ecological households have existed for a long time, most of them private households of broad-minded individuals. Most urban residents have limited time, space and dollars; and they can not envision an environmentally sound lifestyle in the city. Is it possible to establish an ecological household with a "different" standard of living, not a low standard of living?

In 1974 the Farallones Institute, a California environmental organization, decided to create a practical working model of an ecological house and lifestyle. They purchased an old Victorian house on a 6,000 square-foot lot in Berkeley, California. Within 1½ years they completed their prototype and opened it to the public.

The Farallones house shows what a motivated urban family can do in spite of the usual considerations of time, space or money. When all the systems in this house are set up, they take only about eight hours of work each week to maintain, two hours per person for a family of four.

OUTDOORS

The front lawn is alfalfa, not grass. The alfalfa feeds rabbits and chickens raised for eggs and meat. American homes have a combined total of roughly 16 million acres of lawn, the clippings of which are usually thrown away. Grass clippings are excellent mulch for your garden; they need never be tossed into the trash again.

Woodchips from local tree clippings replace cement sidewalks around the house. Concrete kills the microorganisms in soil, whereas wood chips nurture them. Concrete compacts soil and does not absorb rainwater like wood chips, causing drainage problems.

Growing vegetables and fruits is the major gardening focus. Vegetables supply about 20% of the body's nutritional needs and up to 50% of the average grocery bill (because of the high amounts of energy used to grow, fertilize, pick and transport vegetables to the consumer).

An Ecological Home 27

Grains can be grown at relatively lower energy costs and stored and transported more easily. In most urban settings dairy products cannot be made at home due to space limitations and, in many instances, local ordinances.

Berkeley weather allows year round gardening, unlike other parts of the United States. Food also grows in containers in the greenhouse. Container soil is composted from the garden, and plants are watered by recycled household graywater.

A windmill fashioned from old oil drums (Savonius rotor construction) pumps water for a fish pond in the back yard. Fish are used for food. The windmill turns under very low winds that prevail in the house's flatland locale, constantly aerating and filtering the pond water.

Next to the windmill are bee hives which produce about 100 pounds of honey annually. The bees pollinate the trees, flowers and vegetables. A Rodale Energy-cycle grinds wheat, sharpens knives and extracts honey from the combs.

INDOORS

A solar water heater brings water temperature as high as 160°, and in the first year of operation provided 95% of the hot water needs of the household. A small backup electric water heater was used for the few days when the sun was behind clouds. The household spent $1,200 on their solar water heater.

The Farallones household uses a third of the average household water consumption. Water is not wasted and drips and leaks are promptly repaired. Water from the shower, kitchen sink and urinal is piped into the garden through a graywater system. The toilet requires no water; the Swedish-made, Clivus Multrum, dry composting toilet replaces the traditional toilet. In two years human wastes and kitchen scraps are thoroughly composted in this toilet and are usable for the garden around ornamental plants.

A wood-burning Jotul stove is the only heating source. This stove is particularly efficient in converting wood to usable heat, rather than sending most heat up the chimney. A handful of wood burns for a long time in the Jotul. The Farallones house is completely insulated. Window shutters, if they are closed at night, will help retain heat and reduce fuel bills. Kept closed during the day, they will keep the house cool.

Several other systems make the household an ecological example for all members of the urban wilderness. Continuous pilot lights have been eliminated; stoves and pilot lights are ignited for each use with a welder's sparker. The shower head has a flow restrictor to reduce water use. An air convection closet in the kitchen naturally cools vegetables and fruits. A parabolic solar water heater is used for heating tea water.

SUGGESTED READING

Cooper, Ted. *Guide to Bees and Honey.* Emmaus, PA: Rodale Press, 1977. Tools, styles and tactics of beekeeping.

Corbett, Judy; Bainbridge, David; Hofacre, John. *Village Homes: Solar House Designs.* Emmaus, PA: Rodale Press, 1979. A collection of 43 energy-conscious house designs. Solar designs that are within reach of middle-income families. Three types of solar hot water heaters. And much more.

DeKorne, James. *The Survival Greenhouse: An Eco-System Approach to Home Food Production.* Culver City, CA: Peace Press, 1978.

Koestner, Joseph J; Kircher, Ralph, eds. *The Do-It-Yourself Environmental Handbook.* Boston: Little, Brown and Company, 1972. Master checklist for conservation and action guide, from home tips for recycling to how to start an environmental information center.

Organic Gardening, eds. *Build it Yourself Better.* Emmaus, PA: Rodale Press, Inc., 1977. Thick manual of do-it-yourself building projects for the self-sufficient home. Gardening-related projects (stands, tubs, cases, lights, window boxes, tools, compost bins, cold frames, arbors, trellises, irrigation, bird houses), food storage projects (dryers, smokehouses) home improvement projects (concrete, walks, fences, walls, gates, furniture). Highly recommended.

INTEGRATED SYSTEMS

Integral Urban House, Farallones Institute, 1516 5th Street, Berkeley, CA 94710.

New Alchemy Institute, Box 432, Woods Hole, MA 02543.

SOLAR GREENHOUSES:

Helion, Box 4301, Sylmar, CA 91342.

Lark, George. Citizens for Conservation and Solar Development. P.O. Box 49173, Los Angeles, CA 90049. Report on solar heating plans for the greenhouse.

Solar Sustenance, Route 1, Box 197AA, Santa Fe, NM 97501.

Walden Foundation, P.O. Box 5, El Rito, NM 87530.

THE ENERGY-EFFICIENT HOME

Your home can be more ecological by following a few simple guidelines. Use manually operated appliances whenever possible; they save money and resources. Doing chores by hand saves money and helps keep the body muscles active. Turn electrical appliances and lights off when not in use. Think recycle—not trashcan. Whenever possible, fix and mend. The ecological household is occupied by people who realize the value of every item and resource and live accordingly, valuing self-sufficiency above comfort.

LIGHTING

It still seems that electricity is the safest and best way to light our homes. Flourescent bulbs are less expensive and more efficient than incandescent bulbs for lighting a work area. The following chart illustrates the differences between incandescent and flourescent bulbs. Use the formula to compute efficiency.

Incandescent			Flourescent		
Watts	Lumens	Life (in hours)	Watts	Lumens (standard colors)	Life (in hours)
25	235	2500	14	700	9000
40	455	1500	15	870	9000
60	870	1000	20	1300	9000
75	1190	1000	30	2360	18000
100	1750	1000	40	3150	20000
150	2880	1000			
200	4010	1000			

$$\text{EFFICIENCY} = \frac{\text{Lumens}}{\text{Watts}}$$

Every home should have a good supply of candles for emergencies. Inexpensive household utility candles are ideal. A broad assortment of alpine candle lanterns are available at camping and sporting goods stores.

A top-quality flashlight is another important item to have at hand. Nothing is worse than a flashlight that will not work when it is needed. Recently some very ingenious flashlights have been manufactured; most notable are the "pump" flashlights which require no batteries. Several different models are on the market, but they all necessitate continuous pumping; the light stays on only as long as you pump. You will be exercising your hand while you are using this flashlight. Because it is the sudden full-power shock on the bulb (of battery flashlights) that burns out the bulb, these pump flashlights should never wear out.

Because continuous pumping is inconvenient, you may want to have battery-operated flashlights around the house also. In my opinion, the best flashlight one can purchase is the Kel-Lite, which for some time had been sold exclusively to police and fire fighters. This flashlight is so well constructed that the manufacturer gives the original owner a lifetime warranty against damage.

What is so good about the Kel-Lite? The barrel and head are made of a high-tensile-strength aluminum alloy, with a special finish throughout that resists scratching and wear, even under the most extreme conditions. I have seen a Kel-Lite fall two stories from a scaffolding and have only a shallow scratch on the barrel. The flashlight was returned to a distributor, and the customer was given a replacement barrel.

The lens of a Kel-Lite is made of lexan, an unbreakable plastic, that is clear and provides one of the highest-intensity light beams. The Kel-Lite is completely shockproof and water-resistant. Most models have a spare bulb stored at the end of the barrel. These quality flashlights come in a broad array of sizes: from two to seven batteries in length, either C or D batteries, and your choice of either a small, medium or large head. The barrel and head are knurled for easy grip when wet. Kel-Lites cost considerably more than other flashlights, but they are worth the extra price. These flashlights are available wherever uniforms and supplies are available to law enforcement officers, and in most gun shops.

Old hunters and trappers poured grease into an old can and stuck a small rag in the can as a wick. When the wick was lit, the can burned and provided light. This method may be both dirty and unsafe in many urban environments. I mention it only as a possible last resort, emergency method for lighting.

For more extended use, you will probably want quality oil lamps in your self-sufficient household. Glass lanterns are available in many stores and mail order catalogs; many have reflector attachments which greatly increase the light output from a single lantern. The only problem with glass oil lamps is possible breakage if jarred or moved in an earthquake. You can prevent breakage by storing the lamp, wrapping it well, and keeping it just for emergencies.

Metal oil lamps are also available; some can be screwed securely into the wall. Metal lamps can burn regular oil lamp fuel, white gas or olive oil. A similar lamp is the rugged hurricane lamp, designed for ships on the rough seas. These lamps are usually square, all metal, except for unbreakable plastic windows, and are designed to be carried, hung, or set on an object.

COOKING

Cooking in the urban home can be more ecological. Gas stoves are cheaper to maintain than electric stoves, and in heavy winds, earthquakes, etc. electricity usually goes out before gas. Often I have been in a home after all the lights have gone out from storm conditions or an earthquake, and the gas stove continued working. Using heavy-duty cookware, such as stainless steel or cast iron, with tight-fitting lids, will make the most efficient use of your gas stove. Use only a flame large enough to cover the bottom of the pot; a flame licking the sides of the pots is wasting gas. It is a fallacy that larger flames cook food more quickly.

Solar stoves are worth considering for daytime cooking. *The Solar Cookery Book* by Dan and Beth Halacy details solar oven and reflector cooker construction. The Halacy's have been cooking with the sun's power for over twenty years and they include an extensive solar recipe section in their book.

I constructed a solar oven from sheet metal; a simple box with double-thick glass set at a 45° angle on the front of the box. The inside has 3 inches of fiberglass insulation and is covered with foil. The outside of the box has four large reflectors which radiate from all sides of the window. The solar oven has a door in the back for inserting and removing foods. The oven's glass window points toward the sun and the reflectors adjust accordingly. The temperature inside the oven can reach 400°. Cooking times will vary, of course, depending on the intensity of the sun that day and the air temperature.

Another popular do-it-yourself solar cooking device is the aluminum foil, parabola-shaped reflector which points toward the sun. The sun's rays are focused to a point or hot spot at which a cooking rack is located. Food on this rack will cook fairly well, depending again on the intensity of the sun.

Wood stoves, for cooking, save on fuel bills in an ecological manner; however, in many situations a wood stove is illegal or unfeasible. Many wood stoves are available. One of the most popular models is the Ben Franklin stove. If you can't afford a few hundred dollars for a Franklin, you might seriously consider purchasing a kit that converts two 50 gallon drums into an acceptable wood stove, primarily for heating. When the drums eventually wear out, the hardware from the kit can be removed and attached to another pair of 50 gallon drums. Such a kit can be purchased from most camping supply stores and mail order businesses that deal in camping and hunting-related gear.

A small hibachi or camping stove is a great asset for fair-weather cooking and emergency use. They work best with coal or charcoal, but paper and hardwood can be used efficiently.

HOUSEHOLD APPLIANCES

City living and electricity go hand in hand. Yet hardly a day goes by without mention of our vulnerable fuel supplies being cut off. Coal and oil are both considered nonrenewable resources, because unlike trees, which can be planted and regrown, oil and coal do not grow back. Strikes of all kinds, Middle East tensions and uncertainties, and other unforseen events make the cities with their jugular-vein dependency on electricity, highly vulnerable. Each one of us has reason to be alarmed.

Are we to bend passively with the wind and accept a future with no electricity? Is this necessary or in our best interests? Are we to abandon all electrical appliances and all by-products of this technological society? Electricity is neither good nor bad. It is a fantastic tool that can be used for our betterment and growth.

One way to eliminate unnecessary dependence on electricity is to think before you purchase electrical appliances. In my opinion, an electric can opener is an unnecessary luxury. Selfwinding clocks are good to have around, although electric clocks are probably better (as long as there is electricity); they tend to be more efficient and will not stop if you forget to wind them. Look around your home; most modern urban homes have a vast array of electrical appliances that make life oh-so-easy, but are not necessary and waste electricity. We need not discard these items, but we should be prepared to live without them. Phase them out gradually, and do not buy unnecessary electrical appliances in the future. The following charts give you information, comparative shopping and conservative use.

APPLIANCE COMPARISON CHART

ELECTRICAL APPLIANCES

APPLIANCE	ESTIMATED ENERGY USE
Blender	1/200 kwh/use
Clock	2 kwh/month
Clothes Dryer	3 1/3 kwh/use
Clothes Washer	6 2/3 kwh/use
Coffee Maker	1/4 kwh/use
Dishwasher	4 kwh/use
Electric Blanket	
twin	1/2 kwh/night
full	3/4 kwh/night
Freezer (Frostless, 15 cu. ft.)	5 kwh/day
Frying Pan	1/2 kwh/hour
Garbage Disposal	1/100 kwh/use
Hair Dryer	2/5 kwh/hour
Microwave Oven (3 times, 5 min. use/day)	1/3 kwh/day
Mixer (Portable)	1/65 kwh/use
Radio	1/10 kwh/hour
Range and Oven (3 times/day)	5 kwh/day
Record Player	1/10 kwh/hour
Refrigerator	
Manual defrost (10 cu. ft.)	2 kwh/day
Frostless (16 cu. ft.)	8 1/3 kwh/day
Steam Iron	1/3 kwh/hour
T.V.	
Black and White	1/4 kwh/hour
Color	1/3 kwh/hour
Toaster (2 slice)	1/20 kwh/use
Toaster-Oven	1/2 kwh/use
Vacuum Cleaner	1/2 kwh/hour
Waffle Iron	1/3 kwh/use

GAS APPLIANCES

APPLIANCE	ESTIMATED ENERGY USE
Clothes Dryer	
with pilot light	1/3 therm + 1/4 kwh/use
without pilot light	1/6 therm + 1/4 kwh/use
Clothes Washer	1/6 therm + 1/4 kwh/use
Dishwasher	1/6 therm + 1 kwh/use
Range and Oven (3 times/day)	
with pilot light	1/2 therm/day
without pilot light	1/3 therm/day

HEATING

Certain accident patterns are associated with space heaters and heating stoves. Contact with the flame, heating element, or hot surface area can cause your clothing to ignite and result in severe burns. Falling against or touching the exterior surface can also cause severe burn injuries. Explosion of accumulated gas while attempting to light the burner is another problem. An unvented space heater or one which is poorly adjusted and dirty can produce deadly quantities of carbon monoxide. In addition, combustion consumes oxygen in the air; therefore, you need adequate fresh air when you use an unvented heater.

Electrical shock is common, often because the victim was wet and touched the control. Using flammable liquids to "stoke" a fire can ignite and cause an explosion. Using flammable liquids (such as gasoline) in the same room with a heater or stove can create the same hazard.

Most space heaters involved in accidents are gas space heaters, but electric heaters also can cause burns and electrical shocks. The use of improper fuels, such as charcoal in a heating stove, poses special hazards of carbon monoxide poisoning and overheating, especially if the stove is not vented properly.

The United States Consumer Product Safety Commission offers the following suggestions for the selection, safe use and maintenance of space heaters and heating stoves. Buy a heater or stove that can be vented to the outside. Old fashioned or secondhand heaters and stoves may be cracked, improperly adjusted, or improperly vented, producing deadly quantities of carbon monoxide. Be aware that unvented heaters are more likely to produce deadly amounts of carbon monoxide than are vented heaters. Unvented heaters are prohibited in some communities because of the carbon monoxide hazard. Buy a space heater or stove with an adequate guard around hot surfaces or the heating coil to keep children, pets and clothing away from the heat source. If you buy an electric space heater, be sure that it has an automatic switch which cuts off electric power if the heater is tipped over. Be aware that you must keep at least three feet safety clearance on all sides of the space heater. This distance requirement should influence your purchase.

Keep a window partially open if you must use an unvented space heater. Plenty of oxygen is required for proper combustion and the fresh air will help prevent the accumulation of carbon monoxide. Use the proper fuel for each heating device. Don't use coal in a wood burning stove, because it can overheat. Don't use flammable liquid on a wood or coal fire, because it can cause an explosion. Don't use charcoal or polystyrene in a heating stove which is not properly vented, because these fuels can produce deadly amounts of carbon monoxide.

Don't use oil in a kerosene heater or kerosene in an oil heater; do not convert any heater to use another fuel without expert advice.

Keep children away from space heaters and stoves, because they can be burned simply by touching the hot surface. Keep at least three feet clearance in all directions around space heaters or stoves. Don't put a space heater near drapes, furniture, or other flammable materials or near traffic lanes. Pay attention to the fire to maintain proper ventilation and rate of burning. Try to keep a fire at a moderate heat, neither too cool nor too hot. If the fire is too cool, it may permit the accumulation of flammable gases and residues which can explode when reheated by adding more coal or wood. Always use a screen around a stove or space heater which has open flames.

Keep the damper open while the fuel is burning. This will provide for efficient burning and will prevent the accumulation of explosive gases and the leaking of carbon monoxide into the room. Learn how to light a gas space heater properly. If you smell gas, turn off all controls and open a window or door. Don't turn on the gas until you have a match ready and lit. Don't give gas time to accumulate, and if you fail to light a heater on the first try, allow sufficient time for the gas to dissipate before trying again.

Never use flammable liquids (such as gasoline) around a space heater or heating stove. The flammable vapors can flow from one part of the room to another and be ignited by the open flame.

Never leave an unvented space heater or stove on overnight, because deadly quantities of carbon monoxide gas may accumulate and kill you while you are sleeping.

Do not use an extension cord for an electric heater unless it is a heavy duty cord rated for at least fifteen amps. Have an electrician check the wiring in the room if you plan to use a heater wire higher than usual wattage. Never place an electric heater near a bathtub, shower, or sink; avoid the use of a portable electric heater anywhere in a bathroom. Never touch an electric heater if you are wet.

Place a metal sheet under a coal or wood burning stove to protect the floor from live coals or burning wood and from overheating.

Never fill a heater with cold oil because as the oil warms, it expands and could spill and flare. Be aware that mobile homes are smaller than most houses and may have less adequate ventilation, more closely installed electrical and fuel-burning appliances, and more combustible construction. Consequently, the use of a space heater or heating stove can be more hazardous.

Have gas, kerosene, and oil space heaters inspected annually to insure that they are properly adjusted and clean. Have heating stoves inspected once a year to insure that all linings and chimneys are intact, that the stove is properly adjusted, and that it is clean. Check stoves for cracks or faulty legs and hinges. Replace loose or missing guards on electric, gas, oil, and kerosene heaters and on heating stoves. Keep all electrical wiring in electric space heaters in good working condition.

Be sure that all ashes have thoroughly cooled before you dispose of them. If your clothing does catch fire, don't run! Drop down immediately and roll to smother the flames. Teach your children how to react to fire.

Never use a cooking stove as a heater. Never leave the oven door open more than a few minutes at a time because an oven can overheat and cause a fire.

To report a product hazard or a product-related injury, write to the U.S. Consumer Product Safety Commission, Washington, D.C. 20207. In the continental United States, call the toll-free hotline: 800-638-8326.

SUGGESTED READING

Alves, Ronald. *Living with Energy.* New York: Penguin and Viking, 1978. A color-illustrated book that covers the many forms of alternative energies with specific examples of organizations.

Baer, Steve. *Sunspots.* Alburquerque: Zomeworks Corporation, 1975.

Brace Research Institute, Saint Anne de Bellevue 800, Quebec, Canada, February, 1973. "How to Build a Solar Still" and "How to Build a Solar Water Heater."

Brown, Edwin W. "Space Heater Hazards." *New England Journal of Medicine,* Vol. 265, no. 16. (October 19, 1961). pp. 794-5.

Browne, Dan. *Simplified Home Appliances Repair.* New York: Holt, Rinehart and Winston, 1978. How to identify and fix all common malfunctions of major appliances.

Byers, Robert H. *Carbon Monoxide Generation by Space Heaters in Tightly Sealed Rooms.* Georgia Department of Public Health and Georgia Institute of Technology. Final Report to National Institutes of Health, U.S. Public Health Service, October 19, 1950.

Clegg, Peter. *New Low-Cost Sources of Energy for the Home.* Charlotte, VT: Garden Publishing, 1975.

Clews, Henry. *Electric Power from the Wind.* East Holden, ME: Solar Wind Company, 1973.

Consumer Guide eds. *Energy Savers Catalog.* New York: G. P. Putnam and Sons, 1977. A guide to available products for energy-efficient mechanical and electrical equipment.

Daniels, Farrington. *Direct Use of the Sun's Energy.* New York: Ballantine Books, 1974. New Haven, CT: Yale University Press, 1964. Solar energy theories and applications.

Dayton Museum of Natural History. *The Do-It-Yourself Environmental Handbook.* Boston: Little, Brown and Company, 1972. An excellent checklist for home ecology. Make sure your home and your community is doing the best it can to be energy-efficient.

Eccli, Eugene, ed. *Low Cost Energy-Efficient Shelter.* Emmaus: Rodale Press, 1976. Even with conventional buildings there are many ways to make a structure more ecological. Here's the book that tells you how.

Eden, Jerome. *Orgone Energy.* Jericho, NY: Exposition Press, 1972. The power of orgone energy and how it can be used practically is discussed in this book.

The Family Handyman eds. *The Family Handyman Practical Book of Saving Home Energy.* Blue Ridge Summit, PA: Tab Books, 1978. Hundreds of tested, practical ideas to save your energy dollars without sacrificing living comfort.

Gladstone, Bernard. *Guide to Simple Home Repairs.* New York: New York Times Book Company, 1973.

Hackleman, Michael. *The Homebuilt, Wind-Generated Electricity Handbook.* Culver City, CA: Peace Press, 1976.

———. *Wind and Windspinners.* Culver City, CA: Peace Press, 1977. Hackleman's books are by no means general; they provide detailed instructions and applications which are valuable in the fields of alternative energy.

Halacy, Beth and Dan. *The Solar Cookery Book.* Culver City, CA: Peace Press, 1978. An excellent book on solar oven and solar reflector construction with tested solar recipes. Authors have written over seventy books on solar energy.

Heronemus, E. W. *The U.S. Energy Crisis: Some Proposed Gentle Solutions.* Paper presented to local sections of the American Society of Mechanical Engineers and Institute of Electrical and Electronic Engineers, Jan. 12, 1972, West Springfield, MA.

King, Serge V. *Pyramid Energy Handbook.* New York: Warner Books, 1977.

Mazria, Edward. *The Passive Solar Energy Book.* Emmaus, PA: Rodale Press, 1979. This book has transparent sun charts locating the sun's position at any time, during any month, for most United States locations. Step-by-step process for choosing and sizing a system suited to particular needs. Design patterns discuss radiant heating, shading and insulating devices, surface colors, thermal storage walls, window openings, greenhouses and other factors that influence the effectiveness of solar heating.

McKinnon, Gordon P.; Tower, Keith. *Fire Protection Handbook.* Washington, D.C.: National Fire Protection Agency, 1976.

The Mother Earth News Handbook of Homemade Power. New York: Bantom Books, 1974. This book is a compilation of home energy articles that have appeared in the Mother Earth News magazine. Some very practical advice and suggestions on wood, water, wind, solar and methane power.

National Fire Protection Association, 470 Atlantic Avenue, Boston, MA 02210. *Using Coal and Wood Stoves Safely. Chimneys, Fireplaces and Vents. Heat-Producing Appliance Clearances.* These three booklets cover safe installation of stoves, chimney connectors and chimneys. Write to this organization for these booklets and specifications.

Nunn, Richard V. *Saving Home Energy.* Birmingham: Oxmoor House, Inc., 1975. Three hundred ways to cut costs and increase efficiency.

Robinson, Steven; Dubin, Fred S. *The Energy-Efficient Home.* New York: New American Library, 1978. An excellent book on every aspect of home energy efficiency. Has comparative tables, appliance energy usage and many other valuable considerations.

Shelton, Jay; Shapiro, Andrew B. *The Woodburners Encyclopedia.* Waitsfield, VT: Vermont Crossroads Press, 1976. A helpful book in the search for an appropriate stove or furnace for your particular needs; covers all home situations. A good background on wood-burning techniques.

Shurcliff, W. A. *Informal Directory of the Organizations and People Involved in the Solar Heating of Buildings,* 19 Appleton Street, Cambridge, MA, 1975.

Stoner, Carol, ed. *Producing Your Own Power.* Emmaus, PA: Rodale Press, 1974. This is very possibly the best do-it-yourself alternative energy book available. Not a general discussion, but a very specific book, covering wind power, water power, wood power, methane power and solar power. A must for becoming self-sufficient in energy.

Traister, John. *Do-It-Yourselfer's Guide to Modern Energy-Efficient Heating & Cooling Systems.* Blue Ridge Summit, PA: Tab Books, 1977. How to choose, install, maintain, troubleshoot and repair solar, radiant, hot water and hot air systems, and how to cut down on energy costs.

U.S. Consumer Product Safety Commission. *Space Heaters and Wood and Coal Burning Heating Stoves,* No. 34, Washington, D.C. 20207.

Vivian, John. *The New, Improved Wood Heat.* Emmaus, PA: Rodale Press, 1977. Design and construction advice for stoves and chimneys, plus useful discussions of the different types of wood for efficient heating.

Wells, Malcolm; Spetgang, Irwin. *How to Buy Solar Heating . . . with-*

out getting burnt! Emmaus, PA: Rodale Press, 1979. This book takes the mystery out of solar heating for the home. An easy-to-use consumer's guide.

Wik, Ole. *Wood Stoves.* Anchorage: Alaska Northwest Publishing Company, 1977. A complete and practical guide to the log cabin dweller or the city survivor. Covers all aspects of making wood stoves. Illustrated with line drawings and black and white photos.

Woods, Chuck. *Reducing Fuel Costs with Solar Energy.* University of Florida Institute of Food and Agricultural Sciences, Research Report for Fall, 1975.

PRODUCTS AND INFORMATION

Alten Corporation, 2594 Leghorn Street, Mountain View, CA 94043. Do-it-yourself water heating systems and insulation.

Duro Test Corporatoin, 2321 Kennedy Blvd., North Bergen, NJ 07047. Energy saving lighting systems.

Garden Way Laboratories, P.O. Box 66, Charlotte, VT 05455. Collector panels, solar water heaters, do-it-yourself plans.

Killer Watt Corporation, 2639 Yates Avenue, Los Angeles, CA 90040. Energy saving lighting systems.

National Solar Heating and Cooling Information Center, P.O. Box 1607, Rockville, MD 20850.

Solar Energy Products, Inc., 1208 N.W. 8th Avenue, Gainesville, FL 32601.

Sun Ray Solar Oven, 12761 W. Alameda Drive, Lakewood, CO 80228. Solar ovens and reflectors. Offers complete portable solar ovens.

Survival Shop, P.O. Box 42216, Los Angeles, CA 90042. Offers a wide variety of useful survival and ecology-related materials. A good source for the Kel-Lite flashlight.

PUBLICATIONS

EARS—Environmental Action Reprint Service, 2239 East Colfax, Denver, CO 80206. Catalog of alternative energy publications.

SUN Catalog. Solar Usage Now, Inc., Box 306, Bascom, OH 44809. Solar products catalog.

Wind Power Digest, 54468 C.R. 31, Bristol, IN 46507.

ORGANIZATIONS AND PRODUCTS

Aero Power, Warren, VT 05674. Wind generators and systems.

Alternate Consumer Energy Society, 4800 Oak Grove Drive, Pasadena,

CA 91103. Access to alternate energy sources, including low cost hardwares.

Domestic Environmental Alternatives, P.O. Box 92, Hathaway Pines, CA 95233.

Earthmind, Mariposa, CA 95338. Michael Hackleman, Mark Danoff, Vanessa Nauman. Retrofitted wind generators and other energy applications plus numerous publications.

Real Gas & Electric, P.O. Box "A", Guerneville, CA 95446. Wind and other alternative energy systems.

Rodale Resources, Inc., 576 North Street, Emmaus, PA 18409.

Total Environmental Action, Box 47, Harrisville, NH 03450.

BATHROOM WATER CONSERVATION

Various areas of home ecology, such as recycling and gardening, are interrelated in the self-sufficient urban home. An easy and effective place to begin household ecology is in the bathroom. Any individual can have an immediate impact upon household and city water usage, without waiting for government legislation. For those of you who want immediate results, the bathroom is a great place to start.

BATHS AND SHOWERS

Contrary to what many people believe, the shower does not use water and energy more efficiently than a bath. Some people shower quickly, while others bask in the luxury of hot water cascading over their bodies, using great amounts of water in the process.

The bath is actually more efficient than the shower, more healthful and more therapeutic. Bath water can be saved and recycled easier than shower water. Washing a few articles of clothing after each bath will reduce weekly wash loads and save water. I recommend using a biodegradable soap, such as Basic H, in your bath water. This biodegradable liquid detergent leaves your bathtub without a ring and your skin free of that "filmy" feeling. I also recommend a small amount of bath oil or olive oil.

By the time one steps out of a steaming shower, the body begins to sweat. Immediately after a shower, instead of allowing the body to continue sweating and eliminating poisons, we generally dry off quickly. Bath time should not be rushed. One hour is ideal.

Start your bath with two inches of the hottest water possible, to heat the porcelain. Let the water drip into the tub slowly. The sound of water gently dripping is therapeutic to the nervous system and blocks out city noises. In many ways the trickle duplicates the sound you would hear sitting beside a river, and offers time for thinking. Unlike a shower, the slow drip of hot water into the tub avoids draining the hot water in the heater.

Every pore on your body is an excretory organ. A quick shower does not cleanse and keep these organs open. Scrub the entire body with a good, stiff, natural bristle brush. Scrubbing removes dead skin cells, opens the pores and invigorates your body. Focus particularly on the feet, hands, buttocks and neck. Scrub the scalp (not the hair) thoroughly as well.

The hair need never be washed, if the scalp is scrubbed hard in this specific way. Incidentally, a good stiff brush can give an excellent "air bath," if no water is available (on a camping trip or in case of a disaster). Especially if you are unable to bathe, use this technique to keep pores open and "discharging." A vigorous dry brushing over the entire body will leave you feeling as if you have just come out of a bath.

If you stay in the bathtub a long time, you will want to regulate both the water temperature and water level. Rather than pulling the drain, fill plastic milk cartons and containers saved from kitchen use, and put them outside the tub. (I'll tell you what to do with them later.)

REROUTING BATHWATER

Recycle bath water another way, by yourself or with the aid of a friend knowledgeable in plumbing. Go under your house and disconnect the drain pipe that goes from the bath to the city sewer system (usually a pipe 1½" in diameter). Screw the appropriate lengths of pipe onto your bathtub drain so that the water goes out to the garden or fruit trees. If you measure everything before you start, the job should not be major. I have rerouted bathtub water from a second story, by running a pipe out the wall and attaching a hose which ran into a garden. Rerouting expenses are minimal: pipe (which can usually be purchased used), threading (if not already threaded), and labor. Plastic pipe is now available and is easy to use.

Remember to plan the new drain pipe with a slight, but continuous, downward slope until it reaches your garden. Flat spots will slow the water flow and cause a gradual buildup of solids that will eventually clog the pipe. Be sure you use exclusively biodegradable detergent; otherwise you will be doing your plants and garden a great disservice—you may even kill them. Many biodegradable detergents are available. Basic H by Shaklee is phosphate-free, made partly with plants, and acts as plant fertilizer.

FLUSH TOILETS

You can begin more ecological toilet usage without changing your present toilet at all. After you get started, you can explore other possibilities. Did you know that a toilet will work even if it is not hooked up to city water? Surprising, but true. The tank on top of the bowl is a reservoir which stays filled by incoming city water. Water rushes into the bowl when you depress the flush lever and force the bowl's contents away, into the sewer. If the water is turned on, new water flows in and the tank fills up again. What happens if you turn off the water to the toilet?

Take two, one gallon plastic containers with handles, such as bleach containers. With a pair of scissors or a sharp knife, expand the small 1" circular opening to an opening approximately 4" square, but do not cut off or damage the handle. Leave these containers next to the toilet. When you need to flush, fill these containers from the bath water containers. Pouring water from both these gallon containers at once, into the toilet bowl, will provide enough force to flush the toilet. Two gallons is all you need to accomplish what the automatic toilet flush requires five gallons to do.

Bring the water level in the bowl back to normal by pouring another half gallon recycled bath water into the bowl. If you save enough bath water, you may never again need to use city water to flush your toilet.

You may be concerned that your bathroom will become wall-to-wall plastic containers filled with water. Not so. A family of four would have no need for more than about ten full containers at any given time. I have seen shelves hold about 15, half gallon containers. Keep four containers between the sink and the bathtub, and maybe six lined up against the wall, opposite the tub. This may sound like a lot of containers, but they do not get in the way, if everyone follows the system faithfully. You will need to develop your own organization, depending on individual space considerations.

Obviously toilet paper can clog drains, but did you know that a sturdy plastic clothes hamper next to the toilet could eliminate this problem? Put a layer of good soil at the bottom; 20-30 earthworms next; and the toilet paper on top. A sign next to the toilet asks that all guests put their toilet paper into the bin. As the bin fills with toilet paper, the worms devour it from the bottom. Compact the paper lightly at regular intervals, and add more soil and worms, and more water, if necessary. When the bin fills to capacity, bury the contents in the worm farm just outside the kitchen door. This container never sends foul odors into the bathroom or house. The lid is a simple swivel top. The contents provide top-quality food for the worms, which in turn provide top-quality food for the garden.

WATERLESS TOILETS

An alternative for those seeking an ecological lifestyle in the city is the waterless composting toilet, a self-contained toilet that relies on no sewer system and turns all your waste into safe compost. Local ordinances may vary; investigate before making a purchase.

There are several good waterless toilets on the market, all of which claim to be odorless. Some require electricity for fans and heaters, others are completely self-sufficient, designed so that all the feces and urine decompose completely on their own, with the addition of garden soil. Human fertilizer can then be used on fruit trees, ornamentals and nonroot vegetables. (In China human feces is recycled as soil fertilizer.)

Waterless toilets are increasing in popularity throughout the world, because people are realizing that monetary and water waste is unnecessary. Americans alone flush away two trillion gallons of water annually (as of 1978). Human wastes are whisked away into septic tanks, costly sewage treatment plants, and even directly into our rivers, lakes and oceans.

Advocates of waterless toilets prefer to call them "biological recycling units for household use." Waterless toilets operate properly with three basic elements: heat, oxygen and the proper liquid-to-solid ratio. All composting toilets work on the same principle: human wastes and kitchen scraps are deposited into a well-ventilated container; bacteria and mold go to work on them, and in the presence of oxygen, turn "waste" into nutrient-rich humus for the garden. These toilets reduce the amount of waste to an incredibly small amount, and are designed so that humus removal is simple, and with the proper ventilation, free of unpleasant odors.

Oxygen is necessary for microorganisms to do their work. Without oxygen the processes will slow down and the bathroom will smell like an outhouse. A properly installed vent pipe is essential. Some toilets incorporate electric blowers to aid in air circulation.

Proper decomposition of wastes necessitates the correct ratio of liquids to solids. Human and kitchen, liquid and solid wastes contain approximately the right ratio. If the toilet contains an imbalance, add peat moss or water into the toilet.

Ideal aerobic decomposition takes place at 90°-95°. Because bacteria create heat when they convert waste to humus, the ideal temperature can be sustained in the larger permanent chambers, such as the Clivus Multrum and the Toa-Throne. Smaller portable models use backup electric heating coils to keep temperatures near ideal levels. Smaller toilets are also insulated to maintain heat.

Large waterless toilets such as the Clivus Multrum and the Toa-Throne have large digesting chambers below the floor. Smaller waterless toilets are designed to sit on the floor and require no installation other than putting a vent pipe in place. Neither type needs any water supply or sewer-connection plumbing.

TESTING A PORTABLE TOILET

I tested an inexpensive, recreational vehicle toilet in my household. I and two others used this toilet for a two-month period. We used this particular toilet to determine indoor feasibility for regular and emergency use; the possibility of safe, individual bodywaste disposal in large cities; potential usefulness of earthworms in bodywaste disposal; and the effects of the chemical additive on the soil.

The toilet I used was called a portable chemical toilet by the manufacturer. It consisted of a heavy-duty plastic bucket with a handle, a seat, an inner cover that fits over the seat, an outer container for the bucket and an outer top. I chose this toilet because of its simplicity, and because the price is easily affordable.

I used only the waterless toilet during this test period, following the manufacturer's suggestions carefully with each use. We recorded odor, the presence of bugs and any other pertinent information.

All toilet testers agreed that the bathroom hardly smelled from the waterless toilet, except when the toilet was opened. The toilet did not appear to attract insects. At one point I tried wood ashes instead of the chemical. Wood ashes alone seemed OK, and so did the chemical, but wood ashes and the chemical together produced a foul smell. If the toilet produces odors, burn incense or dried herbs, or set out a baking soda and lemon mixture.

I buried the toilet's contents in a large hole covered with ashes, alfalfa straw, goat manure and soil. We also decided to try baking soda to determine its value as a deodorant, and to see how well plants grow in the trench where baking soda is present. Plants in the trench, particularly tomato plants, seemed to thrive and produce prolific amounts of cherry tomatoes for several months. I used the tomatoes, and even served some to friends in salads. I asked many questions to see if anyone experienced ill effects, and no one did. I can't be certain that minute amounts of toxins or slow-acting toxins did not enter the tomatoes. I cannot recommend using fruit grown from toilet trenches, when chemicals have been added to the toilet. Using baking soda exclusively in your waterless toilet, with the possible addition of lemon and lemon peels, you should have no worries.

We conducted our waterless toilet test, not to recommend any specific brand of waterless toilet, but to test the feasibility of waterless toilets in general. I did not feel the particular toilet I used was the best, but it is probably the best in the $15–$25 price range. More expensive RV toilets are available, costing up to $100, and possibly they are worth the money, if you plan to use such a toilet regularly. Some of these small portable toilets can even be flushed and are designed to be virtually free of odor. The toilet would definitely not last a lifetime, but it seemed to be a satisfactory emergency or short-term toilet.

If you do not intend to deviate from the conventional toilet, consider at least putting a full jar of water or sand into the toilet tank to reduce the amount of water for each flush. For emergencies, or just to be closer to nature, have a shovel handy, so you can go out easily to a secluded part of the yard and "get a load off your mind." If you really want to go all the way, use large leaves, such as maple, sycamore, mullein and geranium as toilet paper. Be sure you know how to identify poison oak and poison ivy, before you reach for your wilderness toilet paper.

SUGGESTED READING

Stoner, Carol Hupping. *Goodbye to the Flush Toilet.* Emmaus, PA: Rodale Press. Water-saving alternatives to cesspools, septic tanks and sewers.

Van der Rym, Sim. *The Toilet Papers.* Santa Barbara, CA: Capra Press, 1978. Here's the book that will convince you what a disgrace the flush toilet is. It gives all the practical available alternatives.

WATER CONSERVATION PRODUCTS

Bio-gas of Colorado, Inc., 5620 Kendall Court, Arvado, CO 80002. (Methane).

Ecological Water Products, P.O. Box 509, Dunkirk, NY 14048. Suppliers of water-conserving shower and faucet heads.

Water Control Products/NA, Inc., 1100 Owendale, Suite E, Troy, MI 48084.

WATERLESS TOILET MANUFACTURERS

Biolet Corporation, Box 645, Beatrice, NB 68310. "Biolet."

Bio-Recycler Company, 5308 Emerald Drive, Sykesville, MD 21784.

Clivus Multrum, USA, INC., 14A Eliot Street, Cambridge, MA 02138. "Bio-Loo."

Ecos, Inc., 21 Imrie Road, Boston, MA 02134. "Humus 5."

Enviroscope, Inc., Box 752, Corona del Mar, CA 92625. "Toa-Throne."

Envirovac, 701 Lawton Ave., Beloit, WI 53511. Vacuum waste-treatment systems; use air instead of water.

La Mere Industries, Inc., 702 Main Street, Walworth, WI 53184.

P.E.E. WEE, Inc., Box 108, Grass Valley, CA 95945.

Recreational Ecology Conservation of the United States, Inc., 9800 West Bluemond Road, Milwaukee, WI 53226. "Mulbank."

II
CITY GARDENING

City gardeners with little or no land must develop ingenious techniques and thinking to perfect city gardening. Don't let lack of space stop you from gardening. Recycled containers, some seeds and dirt are all you need to have food growing inside your house, hanging in a patio or sitting on a windowsill. How about starting a rooftop garden to make use of previously wasted space? Just because you don't live in the country does not mean you cannot grow food.

Expand your mental constructs to include variations upon common garden techniques. Even kitchen sprouting is gardening: growing fresh food right in the city. For those who have outdoor gardening space, improve the quality of available soil by using space wisely. Use succession planting as much as possible to increase your garden's yield. Extend the growing season with early plantings, cold frames and flats. Build your own greenhouse. Consider drought-resistant methods and mulching. Do not think limitation—think new possibilities.

THE CONTAINER GARDEN

"Me grow vegetables in containers?? Are you kidding? I live on the fifth floor of an apartment house. I know I haven't enough light to grow these things. And where am I going to get soil around this concrete jungle?" Sound familiar? Well, don't feel alone; many apartment dwellers are in the same predicament.

For lack of garden space, and especially for the apartment dweller, containers suffice. You can grow many fresh herbs and vegetables in containers. You don't know anything about gardening you say? Don't worry; follow these simple guidelines and you'll be eating fresh food from your "garden" in a very short time.

CONTAINERS

Many household containers can be recycled for indoor, patio, and rooftop gardening. New containers are unnecessary. You need not rush out to your local nursery and spend your hard-earned dollars on fancy-looking pots, tubs and containers. As a matter of fact, if you are the average city dweller, you throw away many potential planting containers.

Let's take an inventory of the "refuse" you threw away last week. Did you throw away any yogurt or cottage cheese containers? How about cans (any size or shape), egg cartons, milk cartons, ice cream containers or plastic bags? Did you discard any chipped pottery, such as cups and bowls? Plastic yogurt and cottage cheese containers come in a variety of sizes and are quite suitable for small plants, such as radishes, a carrot or some greens. Treated carefully, these containers can be used and reused many times. If the appearance of radishes growing in a cottage cheese container is unappealing to you, paint the container a pleasant color or cover it with some recycled wrapping paper.

Tin cans of all shapes and sizes are excellent for the container garden. If you have an outdoor garden plot, you can use the cans as "starters" for your seeds before transplanting them outside. Grow green onions in a large can, closely bunched together; thin and eat them as they grow. Grow tomatoes out of a large old can.

Punching drainage holes is unnecessary, if you have the proper soil mix and small rocks or pieces of broken clay pots at the bottom; preparation need not be fussy or messy. If you don't punch drainage holes, you can water less often, but be careful not to overwater, or you will have plants with rotting roots. Water only when the soil becomes dry to the touch. If you decide to add holes for drainage, put them on the vertical face of the can rather than at the bottom. This allows a little bit of moisture to remain in the bottom of the can as a reserve for when you forget to water.

Plants started in tin cans can go directly into the outside ground without removing the can. As the can gradually rusts away, it adds iron to the soil, a nutrient plants can use. Punch a series of holes around the bottom vertical side of the can before planting; the holes allow the roots to grow through the can, while the can protects the young roots against gophers and moles. The can gradually rusts away, but it will last long enough for annual vegetables to mature. After you harvest your spinach or other vegetables from such containers, the container can be added to your worm bed or compost pit.

Chipped cups and bowls, although no longer useful for eating or drinking, make highly attractive containers, especially for herbs. Although egg cartons are not permanent containers for your indoor plants, they are ideal for sprouting seeds. If you use cardboard-based egg cartons (highly recommended over the plastic/styrofoam variety), you can plant the newly sprouted seed directly into the larger container without taking it out of the egg carton. Each of the twelve sections of the cardboard egg carton come apart rather easily when damp and decompose quickly.

Paper ice cream and milk containers make excellent planters. In some cases, you can grow vegetables through to harvest time and never take them out of the containers. A few radishes or one chard plant, for example, could be grown easily in a large ice cream container. You can plant these containers with the plant directly into the ground; they will decompose and help avoid potential root damage that is so common when transplanting many plants.

Undoubtedly many other items can be pressed into service, rather than carelessly tossed away. Choose a container that corresponds to the size of each fully grown plant. For large containers, I suggest you start watching your neighbors' trash cans on collection day. I am continually amazed at the high-quality items even poor people throw away. Once I picked up a one-foot-diameter potted cactus plant that had been put out as trash. Such a cactus would sell for up to $10 at a cactus nursery! I have no idea why it was discarded—possibly the owner no longer cared for it, had no more room for it, or did not want to take the time to see which neighbor would be be delighted to have such a treasure. The cactus had been potted in an old glazed, ceramic cooking pot. Even our old pots can be used for container gardening. I've seen bathtubs, toilets and sinks used as containers for both indoor and outdoor plants.

Sometimes at nurseries 4" plastic containers (or larger) are discarded after plants have been transplanted. In other cases, you might have to pay a small sum for them. Watch the tops of trash cans: people often purchase plants from nurseries, and after transplanting, toss away the plastic containers.

Markets and factories occasionally discard wooden boxes. Keeping your eyes open, you'll have no problem finding containers for your container garden.

"OK, OK. I see now that I can get containers. But believe me, I really don't have enough light to grow vegetables in my apartment." It is unlikely that this person has no suitable location for container plants. Even a small apartment on the darker north side of a building has some light. Most plants prefer as much light as possible. Generally, the southern side of a house (in the northern hemisphere) receives the most light, but containers of certain herbs and vegetables will grow even in the kitchen or bathroom. The kitchen is an ideal place to grow chives and parsley, so they are accessible during cooking. Aloe vera can be grown in the kitchen container garden, easily available to soothe a burn or cut. Humidity makes the bathroom a good place for many plants.

If it is possible in your home or apartment, consider installing window boxes for gardening. A recycled box can grow herbs and small vegetables right outside your kitchen window. Talk to your landlord. See if you can start a rooftop garden. Or maybe there is a plot of land somewhere on the property that you and other tenants could use for a gardening plot. Speak up and ask: you'd be surprised what is available if only you communicate.

SOIL

Gravel, sand, loam and clay are the four basic types of soil. Gravel soils are usually not arable. Sands are coarse and porous, and subject to leaching. Clay (tightly-packed particles of rock) is hard for rootlets to penetrate. Loams, a mixture of sand and clay, are best for gardening. You can get an idea of your soil's makeup by rubbing it between your fingers. Sand is gritty, wet or dry; clay is harsh and hard when dry, and gummy when wet; loamy soil mixes both of these qualities to your plants' advantage.

The apartment dweller can make his/her own batches of potting soil in a large tub or bucket. This should be done in the most convenient place in the apartment (a back porch area, service porch or patio). Begin with a loamy soil and add grass clippings, coffee grounds, fireplace ashes and leaves. Many commonly discarded items can build the soil, such as egg shells and old tea leaves.

PLANTING

Consider a few basic guidelines before putting soil into your containers. First add a layer of small pebbles, broken pieces of clay pots, or other coarse materials to the bottom of the container, whether or not it has drainage holes. Nurseries often sell bags of charcoal bits for the container bottom; they claim it prevents root rot and keeps the soil fresh. If you have a fireplace or wood stove, you need not buy charcoal at all; simply use the small pieces of charred wood that remain after a fire. If you have neither a fireplace nor a wood stove, don't worry—use pebbles. Both pebbles and charcoal/ash material help facilitate drainage in your containers.

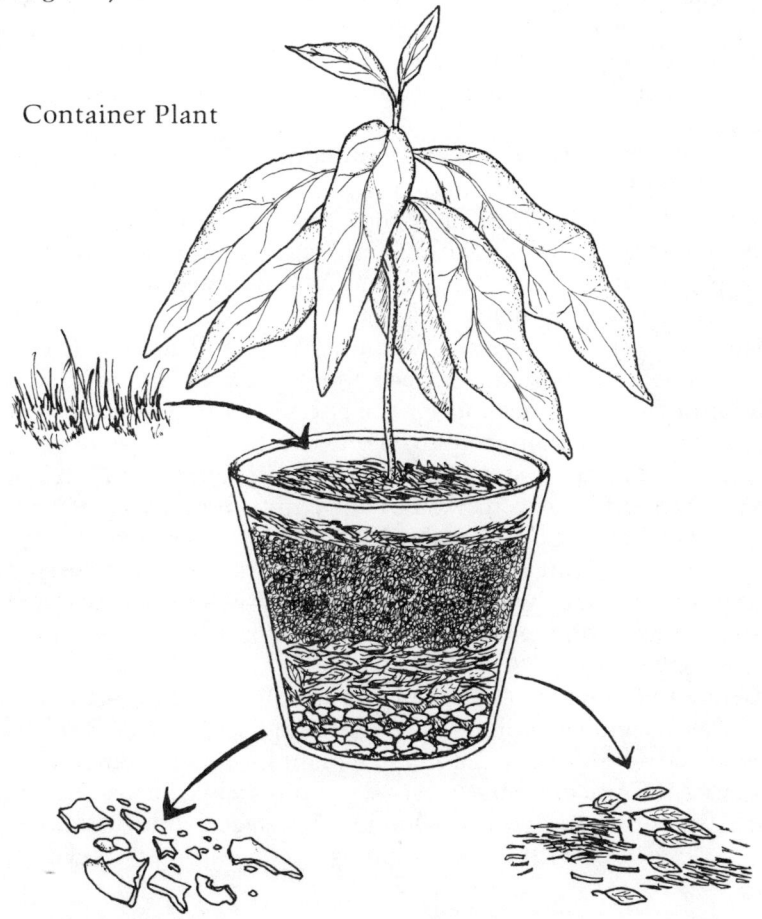

Container Plant

Fill the container with good-quality loamy soil. Do not pack the soil so tightly that all air pockets are compressed. A few firm shakes and gentle pressing over the seeds is sufficient. Remember to leave some space above the soil level to hold water. You may want to put some grass clippings, weeds or leaves from your yard into the pot after you have added the pebbles, before the loam goes in. The organic matter will eventually decompose and provide the plant with a storehouse of nutrients. If grass clippings are available, you can also mulch the top layer of the container with them. Other suitable mulches available to city folks are coffee grounds, dried leaves, and dried and powdered egg shells. The dying older leaves of each plant should be returned to the soil of the particular plant to resupply lost nutrients.

SPECIFIC CONTAINER PLANTS

Tomatoes have generally been kept in containers only long enough to sprout, before they go into the outside garden. Now people everywhere, in apartments, with no yards, can actually grow tomatoes in containers. Hanging tomato plants can be very beautiful in the outdoor patio or porch. Imagine beautiful, red tomatoes on sunny window ledge.

Select plants, such as miniature tomatoes which have been hybridized specifically for container growing. Miniature tomatoes have abundant and compact fruit, and they mature early. Most nurseries and seed companies sell a "Tiny Tim" variety miniature tomato; it grows approximately 15 inches tall, has 3/4 inch tomatoes and can be potted or grown in a porch box.

Miniature tomatoes need a container that will hold at least two quarts enriched soil (tomatoes need food). Begin filling a container with ceramic pieces or rocks; layer the pot with different-sized materials for good drainage. Plant tomatoes one to a pot. Start tomatoes with a large sprout and get a start on the season. Wherever you decide to grow your tomatoes—porch, window or patio—remember that tomatoes need six hours of sun a day. Water tomatoes when they are green and growing, but stop watering when the plant is blossoming (even though the plant will probably wilt and yellow). You want the plant to produce tomatoes, not green bush. Don't be afraid to pick off yellowing leaves.

Many edible and medicinal plants and herbs grow well in containers. Cactus and succulents need little care and many have beautiful flowers. Grow aloe vera as a kitchen plant for its medicinal, gelatinous juice. Aloe juice cleanses, soothes and heals burns (including sunburn) and other irritations. Aloe has 200 species, but Aloe Vera is said to be the best. Filtered sunlight is preferable. Water every two months.

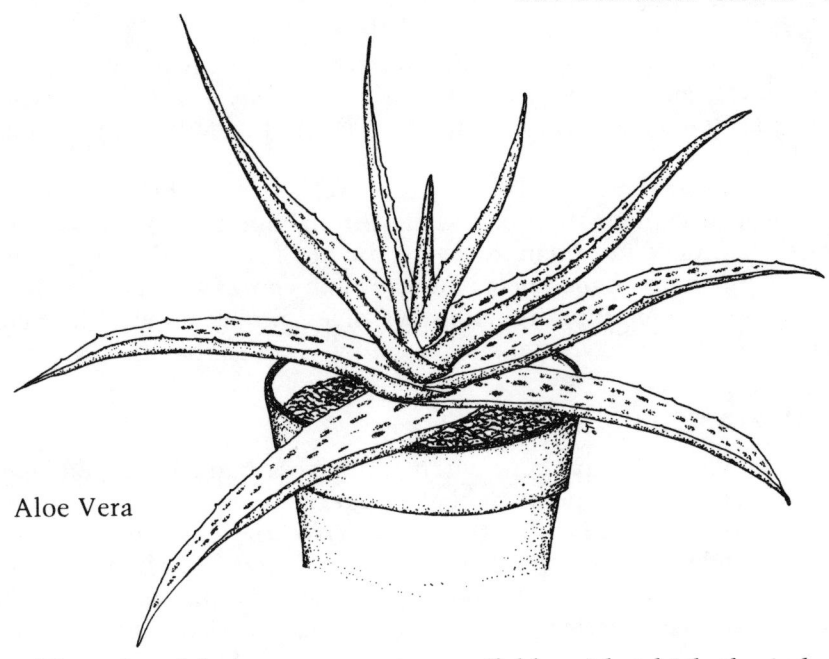

Aloe Vera

Many dwarf fruit trees are now available with which the indoor gardener can experiment. Figs, oranges, lemons, apples; the list goes on and new ones are developed daily. Field's "American Wonder Lemon" produces all year round in sunny windows. This "tree" grows only 2 feet tall, but the fruit tastes the same as regular lemons and grows just as large with fragrant blossoms. The "Brown Turkey" everbearing fig tree from George W. Park Seed Company is designed especially for tubs. Even after freezing weather, this miniature fig tree will go on to produce fruit.

Strawberries are probably the most popular container food around. They will grow in just about anything and look especially beautiful hanging or in window boxes. Mulch your strawberries with pine needles from your Christmas tree for a sweet, wild flavor. Miniature plants make pleasant edgings in small yards and they have concentrated sweetness.

Strawberries are not fussy about soil or climate. They even do well next to other plants and flowers. Give strawberries a moderate, organic-rich soil. Strawberries like a lot of moisture, but they will endure dry spells. Plant strawberries so the crowns are not submerged. Keep the base of the crown at ground level.

Alpine strawberries start easily from seed. The "Quinault" and the "Ogallala" are everbearers, resistant to drought and foliage diseases. The "Ozark" strawberry has a high yield.

Lettuce, Swiss chard, spinach, green peppers, scallions (green onions) and new potatoes are all suited to your indoor or outdoor container garden.

The variety in herbs sets no limitations on what you can grow in containers: anise, parsley, basil, oregano, chives, dill, mints, chervil, tarragon, fennel, marjoram, summer savory, mustard and watercress are only a few.

A number ten can will grow chives for a season of seasoning; you can even dry them. Sprinkle seeds throughout the can or box and water. Thin and use them as they come up. The growth will be thick and lush; what you pick helps more growth come in, and you will not notice any fewer chives remaining. Chives provide an abundant container source of tangy vegetable seasoning as well as healthful greens.

THE HANGING GARDEN

Anyone with a patio or porch can have a hanging vegetable and herb garden. Certain plants will grow out of containers on the ground, on benches, or hanging. Plants can even grow directly in the soil, grow upward and then hang. Train your plants to grow upward, in and out of a trellis, around posts and banisters. The hanging garden is both decorative and practical, and it masters use of that precious urban space.

Choose varieties of plants that are vinelike or can be trained to grow vinelike. This type of garden should require no more attention than any other. Imagine sitting in your patio, shaded by the thick growth of all your vegetables. All you'd need do for dinner is reach up and pick vegetables.

Many plants will grow in this manner. Most members of the gourd family will grow in the hanging garden, including cucumber, smaller squash, zucchini, some species of watermelon, and others. Tomatoes do quite well in the hanging garden, as do peas, passion fruit, strawberries, grapes and all beans, including soy beans and green beans. Some of these grow vinelike naturally and need very little help. Other plants can be carefully pinched and pruned to grow upward rather than outward.

Chayote is another excellent fruit for the hanging garden. Plant the wrinkly, green pear-shaped fruit just as you purchase it from the market. Plant the fruit horizontally in a pot with half the fruit underground. This position closely duplicates the way the fruit would fall naturally from a vine and begin to grow on its own (with no human assistance). Plant two chayote fruit at a time (two are needed for pollination). A vine will develop in time and produce more fruit than you could possibly use yourself. Eat chayote raw or cooked.

Try growing the luffa sponge plant (a type of gourd). You can purchase the seeds from many seed catalogs. When it matures, peel off the outer skin and you will have a durable bathtub "scrubber."

Seed catalogs are an excellent place to check for varieties of plants

which adapt easily to the hanging garden. I've noticed that Lakeland Nurseries' catalog in Hanover, Pennsylvania 17331 has quite a few hanging vegetables. Write to them for their current catalog.

DWARF TOMATO SEED SOURCES

1. Burpee Seed Company, 300 Park Avenue, Warminster, PA 18991. "Small Fry," "Pixie Hybrid."
2. Grace Gardens, Autumn Lane, Hackettstown, NJ 07840. "Sweet 100."
3. J. A. Demonchaux Company, 225 Jackson, Topeka, KA 66603. "Red Cherry."
4. McLaughlins Seeds, P.O. Box 550, Mead, WA 99021. Has many varieties of dwarf tomatoes.
5. W. Atlee Burpee, P.O. Box 748, Riverside, CA 92502. "Pixie Hybrid."
6. Spring Hill Nurseries, 110 West Elm Street, Tipp City, OH 45371. "Toy Boy," "Small Fry."
7. J. W. Jung Seed Company, Randolph, WI 53956. "Tiny Tim," "Small Fry."

8. Henry Field Seed and Nursery Company, Shenandoah, IA 51602. "Tiny Tim."

MIDGET VEGETABLE SOURCES

1. Henry Field Seed and Nursery Company, Shenandoah, IA 51602. "Baby Head Cabbage," "Tom Thumb Lettuce," "Midget Sweet Corn," "Criterion Dwarf Apple," "Baby Cucumber," "Tahitian Orange," "American Wonder Lemon."
2. George W. Park Seed Company, Inc., Greenwood, SC 29647. "Brown Turkey," "Everbearing Fig."

SUGGESTED READING

Ballard, Ernesta D. *Garden in Your House.* New York: Harper & Row, 1971.

Brimer, John Burton. *Growing Herbs in Pots.* New York: Simon and Schuster, 1976.

Elbert, Virginie F. and George A. *Fun Growing Herbs Indoors.* New York: Crown Publishers, Inc., 1974. History of herbs, indoor herb culture, herb gardens in terrariums, flourescent light culture. Step-by-step care for herbs indoors, including growing herbs from seed.

Foster, Gertrude B. *Herbs for Every Garden.* New York: E. P. Dutton & Company, Inc., 1966. How to start a successful herb garden, whether in yard or on windowsill; herbs used medicinally, for culinary purposes or as decoration. Includes garden layouts, how to freeze herbs, how to use herbs to ward off pests.

McDonald, Elvin. *Gardening in Containers.* New York: Grosset and Dunlap, 1975.

Sunset eds. *Gardening in Containers.* Menlo Park, CA: Lane Publishing Company.

Swenson, Allan A. *My Own Herb Garden.* Emmaus, PA: Rodale Press. The manageable kitchen herb garden for children, specifically sighting herbs that are easy to grow.

Taylor, Jean; Davidson, William. *Garden Indoors.* Salem, NH: Hutchinson, 1971.

THE KITCHEN GARDEN

A way to make the kitchen greener and recycle some of your vegetable scraps is to turn them into beautiful, edible houseplants. Some of us have already done this, possibly as a grammar school project. Turn your kitchen into a vegetable garden without going to the nursery.

FRUITS AND VEGETABLES

Cut the top inch off a carrot and place it in a bowl of water. If you keep the bowl from drying out, adding a little water occasionally, you will have a luxuriant ferny plant within weeks. If the greens survive long enough, you can pot them. Pinch off some of the carrot tops for salads, or simply enjoy the beautiful plant.

Avocados are one of the most popular seeds to sprout. They make beautiful houseplants, and within a few years can produce a top-quality food for you and your family. To sprout the avocado seed, cut about 1/4 inch off both the top and the bottom, and peel the seed of its skin. Push tooth picks into the seed so the picks will rest on the jar, and the seed can be suspended in water.

Avocado seed sprouting provides an excellent way to recycle those dark brown and green jars, because the avocado seed needs darkness to be able to sprout. Fill the dark-colored jar with water and suspend the avocado seed over the mouth. The first sprout may take a few weeks to appear, but be patient, and don't let the seed dry out.

As the plant grows it can stay in the jar of water until it is obviously top-heavy. Keep it in the sun while it is growing. After taking it out of the jar, plant it in a large container (even a large old wastebasket will do) where it can grow for several years. If you intend to keep the avocado sprout as a houseplant, you may want to "top" the plant and keep it small. Topping is simply cutting off the top inch or so of the plant, where new growth is appearing. Topping will prevent the central stalk from continuing upward and force the plant to produce lateral branches instead, producing a bushier and more attractive houseplant.

The inner sections of celery, cabbage or lettuce will recycle in very creative ways to produce edible ornamentals. Using celery, cut the head horizontally at the base to release most of the stalks. Peel back the stalks for food and leave only the center core of light white stalks. Place the core in a bowl of water; allow it to grow and turn green.

Peel away lettuce and cabbage leaves from the head, and leave the inner stem along with a few leaves. Place a bowl in a sunny location. Your lettuce and cabbage cores will grow eventually and provide unique, decorative and edible houseplants.

Pineapple is very interesting to grow. Cut off the leaves and top two inches of the pineapple. Plant directly into rich loamy soil. Cover the top of the pineapple with soil, leaving only the leaves above the soil. The pineapple may develop roots and grow. Pineapple makes a beautiful houseplant that is related to the popular bromeliads. Producing fruit from this recycled plant is possible but does not occur as often as we would like. After the top of the pineapple has successfully rooted and shows signs of growing, put half an apple in the pot and cover the entire pot with a plastic bag to help the plant produce fruit.

Plant individual garlic cloves in a pot and allow them to grow to maturity. An individual clove will produce leaves continuously. Pick and eat them like chives. Each clove multiplies into an entire bulb at the end of a growing season. Dig up the bulb and use it, or let it continue growing and pick the healthful greens.

The onions you see in the store that have begun to sprout are excellent for starting your own pot of onion greens. Onion and garlic are perennials; they will continue to produce edible leaves year after year without new plantings, providing you don't dig them up to eat the bulbs. Or start sprouted onions in a jar until they grow larger leaves.

Have you ever grown a potato or yam vine? Children in school do. Just suspend the whole potato or yam in a jar with toothpicks as you did with the avocado. And then, wait. Just make sure the water does not dry out. Eventually a beautiful vine will grow, making a very attractive house plant. When it outgrows the jar, plant it in a larger container.

Save seeds from your fruits and vegetables for sprouting and growing plants. I have seen overripe lemons, oranges and grapefruits with seeds that have already begun to sprout. Have these seeds produce a citrus tree for you. The ideal way to let the seeds grow is to cut the fruit in half, set it in a bowl, and let the already-sprouted seeds obtain their nourishment from the fruit. As the seed becomes larger, it will require a pot and lots of tender loving care. Because citrus like hot weather, be sure to give them plenty of sun and keep them in the warmest part of the house (especially if you live in a very cold part of the country).

Seeds from green peppers, apples, peaches, pears, nectarines, prunes, watermelon, cantaloupe, papaya and other fruits and vegetables can be saved, dried and planted. I prefer to eat some seeds, such as apple seeds (you really can eat the entire apple core), and the inside part of prune seeds. Crack open a prune seed sometime and eat the delicious juice and nut inside.

SPROUTING

Sprouting is a very important and inexpensive way to have fresh, healthful vegetables on hand anywhere, all year round, even in the dead of winter. Sprouts are especially useful in emergency situations, when fresh vegetables are scarce; any beans or seeds will sprout into delicious fresh food. Sprouting is especially helpful to people who cannot afford organic vegetables or who are old and cannot get out often to shop.

Easily digested proteins, vitamins and trace minerals are among the benefits. Sprouted seeds are an excellent source of Vitamin A, B complex, C, D, E, G, K, U, calcium, phosphorous, chlorine, potassium, sodium and silicon in forms the body can assimilate easily. Sprouts have less caloric content than unsprouted or cooked beans or seeds.

Sprouting at home is easy and adventurous. Get your children in on the project. Sprouting is a good way to have children begin helping with food preparation; it will give them self-confidence and early beginnings of self-reliance. Sprouts need ventilation to grow properly. Either purchase a sprouting jar with a screen lid or improvise your own sprouting jar. Reuse old jars, and make cheesecloth tops. Secure the cheesecloth onto the jar with a rubberband.

First soak the seeds overnight in lukewarm water, in a dark place. In the morning drain the water (it can be used for soups or stews). You will notice that the seeds have expanded quite a bit after the long soaking. Keep the jar on its side, so the water can drain, and rinse the seeds or beans at least twice a day. Keep the jar where you will see it and remember to rinse.

The seeds will begin sprouting almost immediately, and within two to five days, you will have fresh, vitamin-rich food to add to salads, sandwiches, breads and cooked foods. Once your sprouts have reached their optimum growth in about five days, keep them refrigerated or they will become tough and turn brown. Sprouts will stay fresh in the refrigerator for over a week.

The most popular sprouting seed is alfalfa. Any beans or grains, such as mung, soy, lentils, garbanzo, wheat or rye will also work. Experiment with any edible seeds, grains or beans for a variety of flavors.

FOOD DRYING

When certain foods are in season, in abundance and sold very cheaply, buy them in bulk and dry them for later. Or if you have a garden producing a large yield of one crop that you cannot use all at once, dry your food. Anything from fruit and vegetables to herbs and meats can be dried. Some dried foods are excellent just as they are, with no further preparation necessary.

Drying food is one of the oldest methods for preserving food. Food can be dried easily in the sun. Stretch cheese cloth over wood frames; slice the food thinly, and spread it out on the cheese cloth; prop the racks up at approximately a 30° angle to the sun.

Foods will require different lengths of time to dry, with air temperature affecting drying time. When drying outdoors, be sure to cover the food with a cloth or screen to protect it from insects. Cover the food at night to prevent dew from settling on it.

Commercial food dryers can cost from $80 to $400. A commercial food dryer usually contains a fan, a vent, a heating device and drying racks. Keep the temperature under 120° to prevent loss of nutrients. Some poor dryers actually cook the food, thereby destroying the nutrients.

I once read about a man who converted his old refrigerator into a food dryer. I decided to do the same. Janice and I took an old refrigerator and removed the motor and rear panel. Everything came out quite easily and neatly. The old-fashioned freezer section was a small box in the top-middle-inside of the refrigerator. We attached a wire inside and hooked up a 75-watt light bulb which we mounted on the bottom inside. Using scrap lumber, Janice made wood frames which fit right where the old metal racks had been.

We dried apples, bananas, tomatoes, zucchini and other foods. Most foods take about two days to dry in my homemade dryer. Food drying at home provides self-sufficiency in food preservation, while eliminating the preservatives of most dried foods, savings in money, and food that will last in any situation, without refrigeration, even in emergencies.

Research of the United States Department of Agriculture indicates that dried food is far more nutritious than either canned or frozen foods.

SUGGESTED READING

Brace Research Institute. *How to Make a Solar Cabinet Drier for Agricultural Produce.* MacDonald College of McGill University, March, 1973.

Collins, J. L. *The Pineapple: Botany, Cultivation and Utilization.* New York: Interscience, 1960.

Dickey, Esther. *Passport to Survival.* Salt Lake City: Bookcraft, Inc., 1969. A very popular and useful book of food storage. Based on the theory that the four musts of food are salt, wheat, powdered milk and honey.

Langer, Richard W. *The After-Dinner Gardening Book.* New York: Collier Books, 1974. Unique guide to growing your own plants from the seeds of all our table scraps. Useful information on soil, cultivation, pruning, and vacation care.

Ruff, Howard J. *Famine and Survival in America.* Alamo, CA: Target Publishers, 1974. This book was influenced by Ruff's reading of *How to Prepare for the Coming Crash* by Robert Preston. Ruff outlines a program of home food storage focussing on nutritional balance.

Saxon, Kurt. *Old Time Home Food Processing.* Eureka, CA: Atlan Formularies. This book gives not only the complete scope of food processing, but also explains how to make a business of home manufacture of processed meats, vegetables, pickles, preserves, catsups, etc. Not a recipe or cookbook.

Seddon, George; Radecka, Helena. *Your Kitchen Garden.* New York: Simon and Schuster, 1978. A complete and beautiful book on every aspect of the kitchen garden.

THE EDIBLE ORNAMENTAL GARDEN

If you are an urban dweller, you probably do not have massive outdoor gardening space. City gardeners often feel they must make the choice between planting vegetables for food, or ornamentals for beauty and aroma (both are valid choices in their own right). Some creative gardeners have found a way to get the best of both worlds with the edible ornamental garden.

To be selected for the edible ornamental garden, a plant must fulfill double roles. Can that beautiful ornamental bush also provide food? If the answer is no, chances are you will not include this bush in your garden. Each vegetable must be attractive enough to provide aesthetics as well as nourishment. Many vegetables are quite beautiful. Rhubarb, chard and asparagus (which produces a beautiful fern with red berries) are excellent examples of food plants that are also highly attractive.

Imagine the differences such a simple shift in thinking would make on the millions of gardens in America. Some ornamentals certainly seem worth time and effort in exchange for breathtaking beauty and heavenly aromas. Gardeners who have seen and smelled these botanical treasures would agree that these plants are well worth having, in spite of the fact that they produce no food and in many cases are poisonous.

Many ornamentals are planted only because they are low-maintenance plants that provide color in an otherwise bland landscape. I not difficult to grow an ornamental garden that also provides food. What is requred is a focus on both food and beauty, when each plant is chosen. Even common food crops can be grown creatively in the ornamental garden. Some examples are rhubarb chard, guava, parsley, currants, mulberry, grapevines and many, many others.

Edible ornamentals include voilets, roses, day lillies, fuchsias and nasturtiums. Violet greens can be eaten either raw or boiled and are a good source of vitamins A and C. Fuchsia is a beautiful plant with a delicate-looking flower, common to many ornamental gardens. Eating fuchsia fruits is not so common, but they are used in making jams and other desserts.

ALYSSUM, SWEET *(Alyssum maritimum)*

Sweet alyssum is another good addition to your salad bowl. This low-growing plant is well known to those in the Western United States, both as a flowering groundcover and a weed. It is a cultivated ornamental plant in the rest of the United States. Sweet alyssum tends to be available year round, although it dies back quite a bit in the heat of late summer and early autumn.

Delicate-looking alyssum has flowers that are typical of its mustard family. The small white flowers are composed of four petals, six stamens (four long and two short), a single pistil in the middle and four (usually green) sepals under the petals. The flowers are arranged in a raceme. Sweet alyssum has a weak and branching stem that lays on the ground and reaches up to a foot in length. The main, somewhat fibrous stem appears to be five-sided in the cross section and is about 1/16 inch thick. The linear to lance-shaped leaves are from ½ to 1½

Sweet Alyssym

Day Lily

inches in length and approximately 1/8 to 3/16 inch in width, tapering to a point at both ends. The leaves are alternately arranged on the stalk, but some appear almost opposite. Although the leaf surface of sweet alyssum appears to be hairless, careful observation through a magnifying glass reveals that both the upper and lower surfaces of the leaves are scattered with fine white hairs. The white flowers of this plant grace gardens, fences, empty lots and roadsides. Pick the entire flower cluster and eat it raw. Adding ¼ to ½ cup of flowers to your salads imparts a mildly hot, watercress-like flavor. Try the leaves and tender, upper portion of the stem in your salads. Because of the hot flavor and the tediousness of gathering the flowers and leaves, sweet alyssum is used mainly as a seasoning in salads. Garnish your meals with a tender flowering sprig of alyssum as an alternate to parsley or watercress.

DAY LILY *(Hemerocallis fulva)*

The most important edible part of day lilies are the peanut-sized underground tubers. Gather the tubers of these plants; be sure to wash and cook them before eating. Boiled or roasted, their potatolike taste is delicious. Unopened day lily flower buds are also a food source, preferably boiled and served as a vegetable. Often day lilies will escape cultivation and grow as weeds.

NASTURTIUM *(Tropaeoleum sp.)*

With their bright orange or yellow flowers and their interesting hexagon-shaped leaves, nasturtiums make beautiful border plants and ground cover. They survive easily with little water, as they have a very shallow root system. Because they like sandy soil, nasturtiums are common to coastal areas and grow noticeably larger there than in other areas. They are perfect for the drought-resistant, low-maintenance garden.

Nasturtium leaves are hot and spicy and liven up any salad or stew with a taste similar to radishes, and the consistency of spinach. Nasturtiums are high in vitamin C and when eaten raw, open the sinuses.

ROSE *(Rosa sp.)*

Roses are common ornamentals, but rarely do you hear talk of people eating them. Roses are a top-quality food source. Dozens of wild varieties and literally thousands of hybrids exist.

The colored part of rose petals can be eaten in a variety of ways. Plucked when the flower bud is in the process of opening, they have a sweet, perfumy taste and a somewhat dry texture, providing a colorful, tasty addition to a green salad. Line the bottom of bread pans with rose

Nasturtium

petals before pouring in the batter. Minced, the petals add subtle flavor to pancake batter, pudding and muffins. Try mixing chopped petals into an omelette. Fresh marjoram goes particularly well with the Rosegg Omelette. Serve a rose petal in the bottom of a glase of wine for visual elegance.

To capture the full flavor, gather the petals in the morning, before the sun shines on them. Snip off the white base of each petal, to avoid possible bitterness. After snipping, you have "prepared petals." Rose petals help make a delicious jam.

ROSE PETAL JAM

1 cup prepared petals
3/4 cup warm (not hot) spring water
juice, 1 large lemon

2½ cups honey (70°-90°)
1 packet powdered pectin
3/4 cup water

Blend thoroughly prepared petals, warm water and lemon at high speed in a blender. Slowly add honey, blending until dissolved. In a separate pan mix pectin and water. Heat until thick. Add water and pectin mixture to rose mixture. Blend together thoroughly and pour into jars. Within six hours at room temperature you should have a jelled product. Use this delicious food on toast, pancakes and in tea.

ROSE HONEY

4 oz. dried red rose petals
3 pints water
5 lbs. mild honey

Cut the white heels off the petals before drying them, as they tend to be bitter. Boil the water and pour over the rose petals; let sit for 6 hours. Strain. Add the liquid to the honey. Boil to a thick consistency.

If the petals remain on the growing stems until they dry, wither and fall off, the round green base of the flower enlarges and ripens into a small but beautiful orange or red fruit. This fruit (or "rose hip") is a rich source of vitamin C, often used in natural vitamin C supplements. One cup of cleaned hips may contain as much vitamin C as twelve dozen oranges.

Before eating fresh rose hips, scrape out the fibrous seeds. Raw hips have a tartness similar to pippin apples. The texture is somewhat similar to a fibrous apple. Rose hips can be eaten raw, added to salads, used in making jams, sauces, soups and teas. Fresh or dried, the hips make an excellent, tart, lemony-tasting tea. Dried, they store better for off-season tea drinking. With or without seeds, rose hips tea is healthful and delicious.

VIOLET *(Viola sp.)*

Many home gardens contain violet plants which produce a beautiful purple blossom in the springtime. How many of us are aware we are planting a good food source when we plant violets?

Pick the young heart-shaped violet leaves for salads or mix them into omelettes and quiches. Don't overdo it with these leaves however, because they are slightly laxative. You may prefer mixing violet leaves with other greens, such as lamb's quarters, chard, dock or spinach, because they are slightly slimy.

SUGGESTED READING

Bryan, John E.; Castle, Coralie. *The Edible Ornamental Garden*. San Francisco: 101 Productions, 1974. This book will change your perspective on gardening and landscaping. It shows how to grow an ornamental garden that you can eat.

Coats, Alice M. *The Treasury of Flowers*. London: Phaidon Press, Ltd., 1975.

Rohde, Eleanor Sinclair. *Rose Recipes from Olden Times*. New York: Dover Publications, 1973. Medicinal virtues of rose petals, hips and leaves; making perfumes, sweet waters, jams, jellies, salads, sauces and confections with roses.

LANDSCAPING WITH HERBS

The traditional front lawn, that beautiful but wasted parcel of land, consumes time, water and energy with little return. Dig it up! Eliminate it altogether, and plant vegetables, or start herbal landscaping.

Outside the kitchen door fill all the extra little pockets of soil with peppermint, sage, parsley, rosemary, thyme, marjoram, chives, possibly dill, and any other herbs that you use in the kitchen on a regular basis. Once you try fresh herbs in the kitchen you will prefer them.

Sundials and herbs need not be restricted to the formal landscape. Plant low-growing thyme around your sundial. On the outside of the walkway surrounding the sundial, plant sweet basil, dwarf lavender and rosemary. Chamomile *(Anthemis nobilis)* will grow in small areas and can even replace the high-maintenance grassy lawn. Give chamomile semishady growing places, along steps or garden paths. It has a low clumplike growth and multiplies. When you want a cup of chamomile tea, just walk outside and pick what you need.

Under shady trees and along paths, grow mints, parsley, sweet woodruff, dill and angelica. Wintergreen *(Gaultheria procumbens)* is low-growing and has pretty red berries. Corsican mint *(Mentha requienne)* is a perfect border plant or ground cover. It has mosslike leaves, is a creeper and repels slugs and snails. Corsican mint will grow anywhere, even between cracks or bricks. Try the wooly, pale-green applemint or bronze orange bergamot mint. Mint likes damp shady crevices.

Yarrow *(Achillea millefolium)* makes a beautiful ferny border. The dwarf variety *(A. tomentosa)* is especially suited to small areas. The gray-leafed santolina has a thick clumplike growth that adapts beautifully to rocky slopes.

For hard-to-garden spaces, between rocks, along stone paths or in the cracks of a brick patio, grow thyme. The thymus species has fifty varieties. Thyme provides a small and neat shrub and has small blossoms. Thyme is a popular culinary herb, medicinal herb and is fragrant.

Do not worry about the shocked neighbors; maybe they will come around to your way of thinking when they see beauty and variation, and smell those fresh herbal aromas. Be the first to bring aroma into your neighborhood.

Landscaping with herbs is rewarding on several levels. Because many herbs are drought-resistant, they can be planted in soil unsuitable to most vegetables. An herb garden or groundcover requires very little maintenance but provides beauty, fragrance, medicine, spice and food. Herbs tend to be more insect-resistant than vegetables, and thus can be planted within the vegetable garden to repel unwanted insects.

PLANTING

When you plan your herb garden, think about the requirements and growth of the individual herb. Perennial herbs are best planted around the edges of the garden or in the flower border, if they are also ornamental. Plant the taller herbs in the background, with lower ones in front near walks and paths.

Herbs may be propagated from seed, rooted cuttings, or division of the mother plant. Starting from seed is best for basil, dill, parsley, sweet marjoram and summer savory. Use division for chives, mints and thyme. Cuttings of sweet marjoram and sage root easily in sand. Mints spread rapidly, and parsley and fennel seed themselves.

Herbs that would do well on sunny slopes include thyme, rosemary, sage, marjoram and oregano. All of these require little care but good drainage. Whether for cooking, tea, bath water, toiletry products, fragrance, or as medicine, these herbs will make an otherwise rocky or drab hillside a beautiful, fragrant, drought-resistant garden.

Line a pathway with thyme, tarragon, chives, and, back from the path, rosemary, lavender and artemisia.

Spilling over a wall, trailing rosemary provides a cascade of blue flowers. The prostratus dwarf *(Rosmarinus officinales)* is especially suited to small spaces. The blue blossoms and pine needlelike, forest green leaves are vigorous anywhere, especially in poor or rocky soil. Rosemary is an aromatic evergreen, drought-resistant and low-maintenance.

The "Munstead Dwarf" lavender *(Lavandula officinales)* will suit your outdoor needs with its gray foliage, purple flowers and aroma. It is a natural insect repellent and moth deterrent.

Lavender is a fragrant hedge suitable for planting along garden pathways. If you have more space, try a border of common thyme with savory behind it. Behind the savory, plant lavender. This is an excellent combination for some patios.

Herbs will grow in containers, along the borders of patios, walkways or porches. Some popular container herbs are marjoram, parsley, chives, nasturtium, summer savory, rosemary, basil and oregano.

HERB GROWING CHART

COMMON, BOTANICAL, AND FAMILY NAMES	LIFE CYCLE PLANT HEIGHT	HOW PROPAGATED
Anise *(Pimpinella anisum)* Umbelliferae	Annual 2 feet	Seed
Basil *(Ocimum basilicum)* Bush Basil *(O. minimum)* Labiatae	Annual 2 feet	Seed
Borage *(Borago officinalis)* Boraginaceae	Annual 2 feet	Seed
Caraway *(Carum carvi)* Umbelliferae	Annual or Biennial 2 feet	Seed
Chervil, or salad chervil *(Anthriscus cerefolium)* Umbelliferae	Annual 1½ feet or more	Seed
Chives *(Allium schoenoprasum)* Liliaceae	Perennial 1 foot	Seed or division
Coriander *(Coriandrum sativum)* Umbelliferae	Annual 3 feet	Seed
Dill *(Anethum graveolens)* Umbelliferae	Annual or Biennial 3 feet	Seed
Fennel *(Foeniculum vulgare)* Umbelliferae	Perennial but often grown as annual 5 feet	Seed

HINTS ON CULTURE	USE AND IMPORTANCE
Sow where plant is to stand.	Seed used as flavoring for bread and cake; leaves in salad or as garnish. Minor importance.
Start in hotbed or plant in open after all frost is past; 1 foot apart.	Young herbage dried for use in soups, stews, salads, and sauces. Important.
Plant in open or transplant seedlings.	Young leaves used in salads; older ones as greens; flowers and leafy tips used in summer drinks; flowers candied. Important.
Sow seed each spring in rows 3 feet apart, with plants 1 foot apart.	Seed used in bread, cake, and cheese; young leaves in salads and soups. Important commercially.
Sow seed in spring or fall (not in summer); thin to 8–12 inches apart; best with some shade.	Leaves used in salads and soups, and for garnishing (like parsley). Minor importance.
Plant any time; best in spring. Divide occasionally.	Leaves used in omelets, salads, soups, etc. Minor importance.
Sow in autumn or spring in 3-foot rows.	Fruits or seeds used in baking and confections; seeds in curry. Minor importance.
Sow in spring; aphids occasionally troublesome if dill is planted late.	Young leaves used in salads; seeds in flavoring, especially in vinegar for pickles. Important.
Plant in spring in 3-foot rows; thin to 12 inches apart.	Seeds used in soups, breads, etc.; leaves for garnishing and in vinegar. Minor importance.

COMMON, BOTANICAL, AND FAMILY NAMES	LIFE CYCLE, PLANT HEIGHT	HOW PROPAGATED
Marjoram, pot (*Majorana onites*)	Perennial but may be grown as annual 2 feet	Seed or division
Marjoram, sweet, annual, or French (*M. hortensis*)	Perennial; grown as an annual 2 feet	Seed or cuttings
Marjoram, wild, winter or Oregano (*Origanum vulgare* or *O. heracleoticum*) Labiatae	Perennial 2½ feet	Seed or division
Mints (*Mentha*) Peppermint (*M. piperita*) Spearmint (*M. spicata*) Water mint (*M. aquatica*) Labiatae	Perennial 3 feet Perennial 2 feet Perennial 2½ feet	Division
Parsley (*Petroselinum crispum*) Umbelliferae	Biennial 2–3 feet	Seed
Rosemary (*Rosemarinus officinalis*) Labiatae	Perennial subshrub 6 feet	Seed or cuttings
Saffron (*Crocus sativus*) Iridaceae	Perennial bulb 10 inches	Bulbs or division

HINTS ON CULTURE	USE AND IMPORTANCE
	Used in soups and stuffings. Unimportant.
Treat as an annual or grow where plants won't winterkill. Protect tender seedlings. Plant 6 inches apart in rows.	Harvest sweet marjoram before blooming; fresh leaves used in salads; fresh or dried leaves used to season meat, cheese, etc. Important.
Divide plants in spring or early fall.	Leaves used in soups, roasts, stews, salad dressings. Minor importance.
	Leaves used in candy, drinks, etc. Minor importance.
Same as for marjoram.	Only commercially important mint. Dried or fresh leaves used for flavoring roasts, soups, potatoes, peas, salad dressing, candy, drinks, jellies, and to give color. Minor importance.
Sow in rows 10 to 12 inches apart; thin to 6 inches apart. Soaking seeds in warm water before planting improves germination. Needs some shade.	Leaves used for garnish and in salads. Popular, but only a few plants are needed in the home garden.
Cuttings root easily; plant in spring in well-drained soil.	Leaves used for flavoring preserves, sweet pickles, jams, sweet bland sauces, meat, soups. One plant is enough for the home garden. Minor importance.
Plant bulbs in early fall, 3-4 inches deep; replant every few years.	Stigmas of flowers dried and used to color butter and cheese; also used to flavor creams, sauces, biscuits, preserves.

COMMON, BOTANICAL, AND FAMILY NAMES	LIFE CYCLE, PLANT HEIGHT	HOW PROPAGATED
Sage or garden sage (*Salvia officinalis*) *Labiatae*	Perennial 1½ feet	Seed or cuttings
Savory, summer (*Satureja hortensis*)	Annual 1½ feet	Seed
Savory, winter (*S. montana*) *Labiatae*	Perennial 15 inches	Seed or cuttings of new growth
Sesame (*Sesamum indicum* or *orientale*) *Pedaliaceae*	Annual 2 feet	Seed
Tarragon (*Artemisia dracunculus*) *Compositae*	Perennial 2 feet	Division
Thyme (*Thymus vulgaris*) *Labiatae*	Perennial prostrate 8 inches	Seed, layers, division, rooted tips

HINTS ON CULTURE	USE AND IMPORTANCE
Root cuttings in early spring 2 feet apart in well-drained soil with moderate amount of soil moisture.	Dried leaves used to season meat, dressings, cheese; young leaves pickled; leaves also used in tea. Important, but one plant is enough for the home garden.
Plant in spring in permanent place.	Leaves and flowers used in salads, stuffings, fresh peas and beans, soups, sauces, rice. Some commercial plantings.
May be planted in borders or in rows 2 feet apart after weather warms up.	Used like summer savory but less important.
Plant in early spring.	Seeds used on breads and rolls and to flavor oil. Minor importance.
Can be grown in poor but not too wet soil; rotate with other crops about every four years.	Leaves and tips used in dressings, tartar sauce, vinegar, preserves, etc.; chopped leaves used in salads, beefsteak, horseradish. Important.
Rooted tips in early spring are the most practical way to propagate thyme; new plants should be started every two or three years.	Dried leaves used in soups, sauces, stuffings, meats, cheese, sour milk, salad dressings. Of some importance.

THE HERBAL HARVEST

Herbs are used both fresh and dry, and their seeds are sometimes harvested for culinary uses. Basil, dill, sage, and thyme may be dried and stored in jars for future use. Fresh leaves of chives, dill, sweet marjoram, parsley, and summer savory are popular. Seeds of anise, caraway, coriander, dill, fennel and sesame are often used.

Herb leaves are best gathered just before the plant blossoms, when the oils are at their peak. Pick the leaves in the morning just after the dew is gone. Snip off the top growth, preferably the top six to eight inches, depending on individual plant size. You need not wash the leaves if they are clean, as many oils may be lost in washing. If the leaves are dusty, rinse them briefly in water.

Many methods exist for drying herbs easily at home. A shady spot is best for leaves, while direct sunlight is best for roots. Try the attic, if you have one, but make sure it is well-ventilated. Spread the herbs on newspaper or cloth, careful that they are not touching each other. If you have an attic, dry herbs there. Avoid drying herbs in rooms of high humidity.

Tying herbs in bundles and hanging them stem upward will keep the precious oils in the leaves and provide aroma and beauty. This method allows the oil in the stems to flow into the leaves. Do not hang herbs above the stove. If you want to avoid an accumulation of dust on your herb bundles, tie a brown paper bag over each bundle with several holes punched in the bag for circulation.

Herbs dry best in three to four days. Don't rush the process. If your conditions are such that the herbs do not dry in a week, dry them in your oven at 100 degrees. Keep them in the oven until the herbs crumble easily. Oven drying is not a good idea, unless absolutely necessary.

Storing herb leaves uncrushed is best, because the leaves will retain more oils. Discard stems before storing. Herbs should be stored in a cool place, out of direct sunlight or strong artificial light. Never store herbs above the stove. Dark glass jars, ceramic objects and various tins make good storage containers for herbs. Do not store herbs in paper or cardboard, as these containers will tend to absorb the oils and leave tasteless herbs.

Shortly after you have stored your herbs, check for moisture. Redry if necessary. Check again in a week's time for mold. If you see mold, you did not dry the herbs thoroughly, and they should be discarded. To use uncrushed, stored herbs simply take out the desired amount, and in one motion, crush and sprinkle them into the food.

MAIL ORDER SOURCES FOR HERB SEEDS, PLANTS AND PRODUCTS

Aphrodisia
28 Carmine Street
New York, NY 10014
herb products

Burpee Seed Company
P.O. Box 748
Riverside, CA 92502
or
P.O. Box 6929
Philadelphia, PA 19132

Caprilands Herb Farm
Silver Street
Coventry, CN 06238

George Park Seed Company, Inc.
Greenwood, SC 29646
herb products

Green Herb Garden
Green, RI 02827
seeds

Gurney Seed and Nursery
Yankton, SD 57078
herb products

The Herb Cottage
The Washington Cathedral
Mount St. Alban
Washington, D.C. 20016
plants

Herb Products Company
11012 Magnolia Blvd.
North Hollywood, CA 91601
herb products

Hilltop Herb Farm
Box 866
Cleveland, TX 77327
plants

Indiana Botanic Gardens, Inc.
Hammond, IN 46325
herb products

Meadowbrook Herb Garden
Wyoming, RI 02898
plants, seeds, herb products

Merry Gardens
Camden, ME 04843
plants

Nature's Herb Company
281 Ellis Street
San Francisco, CA 94102
herb products

Nichol's Garden Nursery
1190 North Pacific Hwy.
Albany, OR 97321
seeds, plants, herb products

Northwestern Processing Company
217 North Broadway
Milwaukee, WI 53202
herb products

Oak Ridge Herb Farm
R.R. 1 Box 461
Alton, IL 62002
seeds, plants, herb products

Penn Herb Company
603 N. 2nd Street
Philadelphia, PA 19123
herb products

Pine Hill Herb Farm
Box 307
Roswell, GA 30075
seeds

Rocky Hollow Herb Farm
Lake Wallkill Road
Sussex, NJ 07461
plants

Snow Line Herb Farm
11846 Fremont Street
Yucaipa, CA 92399
seeds

Sunnybrook Farms Nursery
9448 Mayfield Road
Cherterland, OH 44026

Taylor's Herb Garden
2649 Stringle Ave.
Rosemead, CA 91770
plants

The Tool Shed Herb Nursery
Turkey Hill Road
Salem Center
Purdy Station, NY 10578
plants

Vita Green Farms
P.O. Box 878
Vista, CA 92083
seeds

Waynefield Herbs
837 Cosgrove Street
Port Townsend, WA 98368
seeds, plants, herb products

SUGGESTED READING

Foster, Gertrude. *Herbs for Every Garden.* New York: E.P. Dutton & Company, Inc., 1966. How to start a successful herb garden, whether it is on a windowsill or outside. Complete instructions for growing herbs; material on using herbs in the vegetable garden to ward off pests; cooking hints and recipes. Especially good for the Eastern United States.

Fox, Helen M. *Gardening with Herbs for Flavor and Fragrance.* New York: Dover Publications, 1972.

Hutchens, Alma R. *Indian Herbalogy of North America.* Ontario, Canada: Merco, 1974. A study of Anglo-American, Russian and Oriental literature on Indian medical botany of North America. Highly recommended.

Hylton, William H., ed. *The Rodale Herb Book.* Emmaus, PA: Rodale Press, 1974. How to use, grow, and buy nature's miracle plants. Covers every aspect of herb interest, including the medicinal uses.

Kamm, Minnie W. *Old-Time Herbs for Northern Gardens.* New York: Dover Publications, 1971.

Keller, Mitzie Stuart. *Mysterious Herbs and Roots.* Culver City, CA: Peace Press, 1978.

Kloss, Jethro. *Back to Eden.* Riverside, CA: Lifeline Books, 1973. A classic herbal. 700 pages. Highly recommended.

Law, Dr. Donald. *The Concise Herbal Encyclopedia.* New York: St. Martin's Press, 1973. An excellent and beautiful herb book.

Lighthall, J. I. *The Indian Folk Medicine Guide.* Norman, OK: University of Oklahoma Press, 1969. A useful book for your herbal library. Unique perspectives.

Loewenfeld, Claire; Back, Philippa. *The Complete Book of Herbs and Spices.* Boston: Little, Brown and Company, 1974.

Lucas, Richard. *Common & Uncommon Uses of Herbs for Healthful Living.* New York: Arco Publishing Company, 1972. A very useful little, mass market paperback; a fascinating account of herbs and their remarkable healing properties as used through the ages. Includes healing plants from the sea, herbs for bathing and beauty, herbal smoking substitutes, folk remedies and much more.

Messengue, Maurice. *Of Men and Plants.* New York: Bantam Books, 1974. This is more or less an autobiography of Messengue and his work as an herbalist. An inspiring work—a must for anyone who doubts the authenticity of herbal healing. Recipes included.

Meyer, Clarence. *The Herbalist.* Glenwood, IL: Meyerbooks, 1976. Medicinal plants, vitamins and minerals, cosmetics, dentifrices, gargles and much more.

Morton, Julia F. *Herbs and Spices.* New York: Western Publishing Company, 1976. 372 species of flavoring plants, both wild and cultivated. Discusses little-known savory plants that give distinctive flavors and flavor enhancers. Special plant parts for seasoning.

Patterson, Bryce L. *Growing Vegetables and Herbs.* Richmond, CA: Brombacher Books, 1975. Helpful hints for the home gardener, whether home owner or apartment renter. How to select a place to grow things, how to prepare it for seeds or seedlings. Up-to-date information.

Prenis, John. *Herb Grower's Guide.* Philadelphia: Running Press, 1974. 25 fascinating and useful herbs that can be grown indoors and outdoors. Even if you have never grown anything before, you can grow the plants described in this book.

Sunset Magazine eds. *How to Grow Herbs.* Menlo Park, CA: Lane Books, 1972. A thorough book on home gardening of herbs and their use in the landscape.

Thompson, Robert. *Natural Medicine.* New York: McGraw-Hill Book Company, 1978.

Walker, Elizabeth. *Making Things with Herbs.* New Canaan, CT: Keats Publishing Company, Inc., 1977.

Wilder, Louise Beebee. *The Fragrant Garden.* New York: Dover Publications, 1974.

THE DROUGHT-RESISTANT GARDEN

Drought-resistant gardening is far more than a way to garden; it is an integrated system of activities, a lifestyle which can enable us to provide our own food and survive periods of drought and famine upon the Earth's surface. Living this lifestyle necessitates having knowledge about water, soil, plants, sun, economics, plus various techniques and devices.

Without water most plants will die. Plants exist, however, which need only minimal water to live and produce food. Knowing which plants is especially desirable for inhabitants of desert areas, places that are dry in the summer, people with little water, and all of us, should drought conditions arise. Rain harvesting is another aspect of the drought-resistant lifestyle. When water is falling freely from the sky, capture it for later use. Rainwater can be collected and stored either directly, using large plastic trash containers, or by diverting roof drainpipe water, through a filter, into storage containers.

CONSERVING WATER

How do we prepare for water rationing? Switching to a nonphosphate biodegradable soap and rerouting the sink and bath makes the water suitable for your fruit trees and garden. With minimal expense rerouting water from the house to the garden will enable you to keep all the water you use right in your own yard. Obviously this method saves greatly on additional garden water and works very well. Water your garden between midnight and five a.m., approximately (during the dew), when it will evaporate less.

During rain the sky tends to be dark and the water comes from above, finding a natural path to the roots. We can simulate rain by nighttime overhead watering. Keep plants clean as well as watered—they too must breathe. Overhead watering is not necessary on a regular basis. Leaving the hose in the garden for approximately one hour with light pressure will give the soil a good soaking; do it about once a week. Place the hose on a board or some newspapers so the soil won't be dug up. Frequent light waterings evaporate quickly, encourage shallow roots, and weaken a plant's drought-resistancy.

The toilet is a major water user in the household. During drought, flush only when absolutely necessary. Add plastic bottles full of water into the tank, so the toilet uses less water with each flush. Are you aware that waterless toilets are available? A good waterless toilet is odorless and requires no water at all. It is a closed decomposition system wherein the feces and urine (and even kitchen scraps) compost into a high-quality fertilizer for your garden.

Wash your car with a high-quality, chemically treated, deep-nap cotton cloth that "washes" without water. Water, contrary to what we have all been led to believe, is the worst thing for a car's paint job. And all that saved water can help grow alot of food. The best auto drywash cloth available is called the KozaK cloth, manufactured in New York since 1927. These cloths can be purchased from the Waterless Car Wash Co., Box 42216, Los Angeles, CA 90042 for $5.50 postage paid (as of 1979).

You may choose to do container gardening. With containers you can garden drought-resistantly even if you have no yard. Be sure not to put drainage holes on the bottom of the containers, but if you do, the bottom of the container should be lined with pebbles and filled with a mixture of soil, compost and worms. Plants in containers require less water, because unless the container has drainage holes, all the water stays in the container. Use old coffee cans and buckets; any recyclable container that will do the job.

Of course there are many other aspects of the drought-resistant lifestyle. The truly observant reader will have noticed that such a lifestyle contains real solutions to the problems facing humanity today, particularly in large cities.

MULCHING

Much more water is absorbed and remains in a mulched garden than in an unmulched garden. A mulch is an organic material that is placed on the surface of the ground around and underneath plants. A vital part of any drought-resistant garden, mulch protects the soil, holds in moisture, reduces the growth of undesirable plants and adds nutrients. You can mulch with leaves, wood chips, rocks, grass clippings, newspaper, straw and any plant material. An unmulched garden needs more water, more work, and will be more susceptible to erosion and caking.

Onion and garlic skins should be buried in your onion and garlic patches. Citrus peels should be dried in the oven or sun, ground up, and used as a mulch under citrus trees. If the clippings from each tree or plant are left where they fall, they provide a natural mulch, a steady trickle of nutrients to the plant. Forests operate in this manner, constantly renewing and feeding the growth. Raking clean under your garden plants is foolhardy and in opposition to the laws of nature.

The idea that all the space between your garden foods must be kept clear of competitive plants or "weeds" is erroneous. Many garden weeds are edible, vitamin-rich foods, such as chickweed, lamb's quarter, sow thistle, dandelion and purslane.

Horse manure can be worked into the soil. Neighbors' grass clipplings can be collected and used as a mulch. Leaves are excellent as mulch or compost. Any weeds picked from a section of the yard should not be discarded; they must be composted.

STONE MULCHING

Stone mulching is an excellent gardening technique for combating drought conditions or for just plain water conservation. Stones have the ability to hold the sun's heat for much longer than other mulching materials. At night, the heat is gradually released, raising the colder temperatures that surround the plants at night. The moisture content and soil temperature remain more uniform in rock-mulched soil.

Stone mulching began when ancient farmers observed that plants growing closest to rocks held up better to adverse weather conditions, matured earlier, continued growing longer in the season and generally thrived, when other plants did poorly.

You might be wondering how anything can grow in a garden covered with stones, but you need not cover the entire garden with stones. Condition the soil, and line both sides of your planting rows with stones. I recommend oak leaves, pine needles, straw and sawdust. Set the stones in place with the flattest side on the top. Fill open areas between the stones with gravel, sawdust or soil.

Many gardeners have reported great success with stone mulching. Stone mulching suits roses, fruit trees and shrubby plants as well as vegetables. During a drought, fruit trees grow substantially larger fruit when stones are set between two thick layers of straw (straw under and above the stones).

COMPOSTING

You can enrich your soil with all your kitchen scraps by composting them. To compost, organic material is either piled or put in a pit in layers and covered with plastic. Once the decomposition process is complete (the time will vary), you will have top-quality fertilizer, the basis of soil fertility. You can never have too much compost in your garden. The best compost (and safest in the city) is made by piling layers (starting from the bottom) of paper and cardboard, tough organic material, kitchen scraps, manure, blood meal, grass clippings, leaves, wood ashes and rock phosphate, and soil with worms.

HEARTY PLANTS

Good soil retains moisture and is more drought-resistant. Some foods prefer drier, less fertile soil. The Toyon tree (edible berries), black walnut, mission fig, mission grape, mission olive, pomegranate, citrus fruits, rosemary, yuccas (edible fruits), oaks (edible acorns), agaves (several edible parts), carob (edible pods), honey mesquite (edible pods), jujube (edible fruits), jojoba (edible nuts), several varieties of cactus (edible fruits and pads), pinyon pine (edible nuts), elderberry, California bay (leaves for seasoning and tea) are good drought resisters.

The Hopi, Navajo and Pueblo Indians of the arid Southwest grew food in the desert with as little as five inches of rain annually. Some of the plants they grew include pima squash, blue maize, devil's claw, teparies and other desert beans. All of the above plants grow naturally in the desert, can survive extended drought, and love the heat. The drought-resistant lifestyle necessitates an expanded awareness of the plants around us.

SUGGESTED READING

Crispo, Dorothy. *The Story of Our Fruits and Vegetables.* New York: Dorex House, 1968. This is a fascinating account of the history of all our common fruits and vegetables.

Fryer, Lee; Simmons, Dick. *Ecological Gardening for Home Foods.* New York: Mason/Charter, 1975.

Fukuoka, Masanobu. *The One-Straw Revolution.* Emmaus, PA: Rodale Press, 1979. An introduction to a unique natural farming method which requires no machines, no chemicals and very little weeding.

Kelway, Christine. *Gardening on Sandy Soil for Northern Temperate Areas.* New York: Dover Publications, 1975.

Langer, Richard W. *Grow It!* New York: Saturday Review Press, 1972. If I had to choose one comprehensive book on gardening and animal raising, this is the one.

Legget, Arnold; Falge, Pat. *The Complete Garden.* Willits, CA: Oliver Press, Charles Scribner's Sons, 1975.

MacLatchie, Sharon. *Gardening With Kids.* Emmaus, PA: Rodale Press, 1977. Alerts adults to what children expect and the best methods for success with them. Includes recipes to show children what they can do with the food they grow.

Medsger, Oliver Perry. *Edible Wild Plants.* New York: Collier Books, 1966.

Minnich, Jerry; Hunt, Marjorie. *The Rodale Guide to Composting.* Emmaus, PA: Rodale Press, 1979. How to custom-make compost for individual soil problems: equipment, materials and techniques. Where to make and use compost.

Nicholls, Richard. *The Handmade Greenhouse from Windowsill to Backyard.* Philadelphia: Running Press, 1975. Materials, construction, climate control.

Olkowski, Helga and William. *The City People's Book of Raising Food.* Emmaus, PA: Rodale Press, 1975. What do you do if you live in the city and have little space? This book gives you plenty of gardening suggestions.

Organic Gardening and Farming Magazine eds. *The Organic Way to Mulching.* Emmaus, PA: Rodale Press, 1972.

Rodale, Robert. *The Basic Book of Organic Gardening.* New York: Ballantine Books, 1971.

Robbins, Ann Roe. *Twenty-five Vegetables Anyone Can Grow.* New York: Dover Publications, 1974.

Seymour, John. *The Self-Sufficient Gardener.* New York: Doubleday & Company, Inc., 1979. How to grow more food in less space with the new deep bed method.

Shuttlesworth, Dorothy E. *The Hidden Magic of Seeds.* Emmaus, PA: Rodale Press, 1976. A book for children, ages 4–7. Tells how seeds are made, how they travel, what sizes and shapes, why they grow. Easy-growing flowers and foods are emphasized.

Sunset Books and Sunset Magazine eds. *Sunset Basic Gardening Illustrated.* Menlo Park, CA: Lane Books, 1971. Detailed gardening basics.

Swenson, Allan A. *Landscape You Can Eat.* New York: David McKay Company, Inc., 1977. A complete guide to planting, growing and enjoying trees, fruits, berries and nuts in your own backyard.

Wheatly, Margaret Tipton. *Successful Gardening with Limited Water.* Santa Barbara, CA: Woodbridge Press Publishing Company, 1978.

WORMS, SNAILS AND OTHER "PESTS"

Unfortunately we have very negative thinking when it comes to worms, snails, ants and insects in general. The time has come to change our attitudes. These detested creatures are viable, healthful food sources; only our educated prejudices stand in the way of using them for food. People are quick to complain about world food shortages, but slow to find solutions.

Balayem, a member of the Tassaday tribe in a Philippine rain forest, speaks about his people's food in the August, 1972 *National Geographic*. "Our streams gave us all else that we needed: tadpoles, frogs, crabs, little fish. We killed no forest animals. We had no traps or hunting weapons. Our ancestors were friends of the deer and could touch them. We also found berries and flowers that are good to eat. One red flower and one yellow one. And wild bananas, and grubs that live in rotten logs."

In a world where famine breathes down our necks, how can we continue to ignore the multiple sources of free, nutritious and delicious food?

WORMS

Worms may be the solution to the world food crisis and the solid waste disposal problem in the cities. Earthworms are valuable because they improve garden and farm soil, convert trash into fertilizer, and they may one day supplement the world's food supply.

Because worms don't like sand or dirt to touch their bodies, they burrow continuously. The burrows are lined with whatever organic material worms find at the ground's surface. The constant making and abandoning of burrows aerates the soil. As the worms come to the surface, they deposit their castings (excrement) which are rich in nitrogen and phosphate. Worms burrow vertically, with a constant plowing that eventually undermines big pebbles and rocks that get in their way.

The roots of plants grown in worm-burrowed soil will be able to move easily through the aerated soil in search of food. Plants and vegetables grown in worm-rich soil grow bigger and better, which produces more surface food for the worms, which produces more natural mulch matter (and on and on).

Water retention in worm-burrowed soil is much greater than in ordinary soil. Aerated, worm-burrowed soil absorbs water like a sponge and helps prevent run-off and erosion. Raising worms to enrich large areas of unusable land is actually practical, because worms reproduce so quickly. Worms double in population every 75 to 90 days, increasing in geometric progression.

One plan on the international scene is to use 32,000 tons of worms to help turn vast areas of Middle Eastern desert into rich farming soil. A British consulting engineer, Mr. Derek Hardingham, invented a machine for inserting a layer of polyurethane foam, 18 to 24 inches below the surface of the ground, as an artificial water table. Acting like a sponge, the polyurethane would stop moisture from seeping back too fast into the desert. After the polyurethane foam is in place, Mr. Hardingham intends to innoculate 115,000 acres of desert with red hybrid earthworms, because they deposit the nutrients necessary for farming.

The process of making a worm bed is relatively simple. Worm beds can be simple four-by-eight foot rectangular wooden frames, about one foot deep. Add compost, a layer of limestone, horse manure, and finally worms. Regularly add coffee grounds, kitchen compost, vacuum cleaner dust and manure to your worm bed. If you use recycled wood, your only expense will be the worms. Ordinary garden worms are unacceptable for worm beds. Growers use the red worm *(Lubricus Rubellus)* because they are the most prolific eaters and breeders, and are able to withstand greater temperature variation.

First fill the bottom of the beds with compost or other organic matter for the worms to eat. Newspaper, cardboard, rags, branches, wood and almost all the "junk" we haul away to landfills every day (with the exception of rubber, glass and metal) can be shredded into worm food.

That worms reproduce so rapidly and eat their equivalent weight every 24 hours makes worm beds a practical method for dealing with the staggering amount of daily waste from our cities. An average worm bed has about 100,000 worms. Imagine thousands of worm beds whose populations double every 2½ months. Instead of filling in our beautiful canyons, sanitation departments could conceivably take the weekly pickup to worm farms rather than landfills. The potential for waste disposal is fantastic.

Earthworms are a high-quality food for humans, and if we can change our prejudices over what is proper food, worm raising could realistically help feed people. Seventy percent of the worm's dried weight is protein. Since eating worms requires an "acquired taste," North American Bait Farms, Inc., in California has been sponsoring annual cooking contests to stimulate public interest and acceptance of worm eating. Some of the more popular recipes are Earthworm Patties Supreme, Applesauce Surprise Cake, Ver de Terre (earthworm) Stuffed Peppers.

Here's a simple recipe you can try at home. Let me know how you like it.

NATURAL TREAT

1½ pounds earthworms
½ large onion, chopped
¼ cup chicken bouillon
1 cup sour cream

3 Tbsp. butter
½ cup mushrooms (optional)
Whole wheat flour

Thoroughly wash and purge the earthworms before using them. To purge, boil the worms three times and then bake them in the oven at 350° for 15 minutes.

Coat the worms with flour and brown them in butter. Add salt to taste. Add bouillon and simmer for 30 minutes, stirring occasionally. Sauté onions and mushrooms separately in butter. Add the onions and mushrooms to the earthworms. Add sour cream. Serve over rice or chow mein noodles. Serves four.

SNAILS

Other garden creatures are also good food sources. The edible European brown snail *(Helix aspersa)*, the French "escargot," is the same common garden snail we consider a pest in our gardens and yards. Known in France also as the *Petit Gris* or vineyard snail, it was first brought to the North American continent about 1883. Another edible snail, preferred by many people from Southern Europe, and common in some parts of California, is the white Spanish or "milk snail" *(Otala lactea)*.

Throughout the world people have been eating snails for centuries. Edible varieties are often cultivated in cages. Their distribution in the United States was from either accidental introduction, such as escape from cages, or intentional introduction by individual people. Italian-Americans in the San Francisco area have enjoyed their local snails, *Bobaluchias*, but few Americans, other than those of European descent, eat snails.

Snails belong to the same family as abalone, the sea snail widely acclaimed as a specialty food. Escargots in the United States are still a garden nuisance to most people, even though they can be prepared as a tasty food, low in calories (about 90 calories per 100 grams of meat), high in protein (12–16 percent), and rich in minerals. Rather than preparing snails at home, however, many people who eat them prefer to order them in exclusive restaurants at high prices. For those willing to try economical, home-prepared escargots, I submit recipes for you to try.

Rather than poisoning our garden snails, we could use them as food. Once again, our mindsets are all that stand in our way. We would save money on protein foods, if we ate snails. The brown garden snail is mature when it is about 1 to 1½ inches long and ½ inch wide. Mature snails are best for cooking, because they have more and tastier meat, and their shells are easier to remove; immature snail shells are thin and difficult to remove. Where snails are collected for eating, use no poisons for snail control. If you have used poison, do not collect snails until at least six weeks after stopping the poison.

Because snails are nocturnal creatures, collection is best about two hours after dark. Take only moving snails. Lightly water the collection area in late afternoon to bring the snails out of hiding. During dry weather they will seal themselves to any surface available.

To use snails for eating, you must first purge them of any off flavor or toxic material from the food they have eaten. Place about ½ inch of damp corn meal at the bottom of a plastic, glass, cast iron or ceramic container. Put snails in the container and cover with a ventilated top. A wire refrigerator shelf, hardware cloth, cheese cloth or nylon netting will provide plenty of air and still allow you to observe the activity of the snails. Weight the cover with bricks or tie it down securely so the snails do not escape.

Place the container in a cool, shady area. Allow the snails to purge themselves by eating the cornmeal for at least 72 hours. Snails will stay alive and healthy in containers for a long time. Replace the cornmeal every other day to prevent it from molding and souring. The snails will feed and crawl up the side of the container to rest. Use only active snails in cooking. After at least 72 hours remove the snails from the container and wash them thoroughly in cold running water, to remove the cornmeal from their shells. They are ready for blanching.

Plunge the snails into boiling water and simmer for about fifteen minutes—the same as you would live shrimp, lobster, crab or crawfish. A bay leaf in the cooking water will give this operation a pleasant aroma. As the water will foam, control the heat to prevent the pot from boiling over. After blanching, drain the water from the pot. Using a toothpick, nutpick or narrow-pointed knife, pull the snail meat from the shell. Save some shells for later use. Remove and discard the dark-colored gall (about ½ inch long) from the tail end, where the snail is attached to the shell. Wash the snail meat thoroughly under cold water. After purging and blanching the snails, prepare according to your favorite recipe or package and freeze for future use.

Boil the empty shells for about thirty minutes in water, adding ½ teaspoon of baking powder per pint of water. Drain the shells, wash them thoroughly in cold running water and dry them. Use these shells to serve snails, in recipes that call for cleaned shells.

SNAILS IN TOMATO SAUCE

1 medium onion, chopped
2 cloves garlic, chopped
½ cup chopped bell pepper
2 Tbsp. oil
1 can tomatoes or approximately one pound fresh tomatoes
1 pint cleaned and blanched snails
Salt and pepper to taste

Sauté onion, garlic and bell pepper in oil. Add the tomatoes, salt, pepper, and simmer until bell pepper is tender and flavors are blended. Add the snails and simmer for ten minutes. Serve hot over hot toast wedges, hot rice or noodles. Serves 4 to 6.

SNAILS IN GARLIC BUTTER

18 cleaned and blanched snails
1 cup water or dry white table wine
Seasoning as desired, such as onion, garlic, bay leaf and salt
18 washed and boiled snail shells
Garlic butter
Grated swiss cheese (optional)
2 Tbsp. fine, dry bread crumbs

Simmer cleaned and blanched snail meat for 10 minutes in water or wine. Season as desired. Drain liquid and save for later. Place a small amount of garlic butter in each empty shell. Place snail meat into each shell and sprinkle with grated cheese, if desired. Seal shell opening with a thick coating of garlic butter and sprinkle with bread crumbs.

Place stuffed shells in a shallow pan containing about 2 tablespoons of the saved snail meat liquid. Bake in a 350° oven for about 10 minutes or until the butter bubbles and the bread crumbs are browned. Serve immediately. Serves three.

Stuffing and refrigerating snails several hours prior to baking is fine. You may also package them in moisture-proof containers and freeze them.

BAKED SNAILS

1 cup snail meat, cleaned and blanched
1 cup water
¼ onion, diced
4 cloves garlic, chopped
Dash allspice
2 bay leaves
2 Tbsp. melted butter
½ – ¾ cup bread crumbs
Dash pepper
Dash salt

Simmer snail meat for 10 minutes in water seasoned with onion, garlic, allspice and bay leaf. Toss the cooked snail meat in melted butter and roll in bread crumbs seasoned with pepper, salt and garlic powder. Place in a shallow greased pan and bake in oven at 350° until brown. Sprinkle with lemon juice before serving. Serves two.

STUFFED SNAIL SHELLS

2 dozen snails, shelled, cleaned and blanched
1 cup salted water
6 cloves garlic, minced
4 Tbsp. olive oil
2 Tbsp. butter

Simmer snail meat in salted water until tender. Chop the snail meat, mix with minced garlic, and sauté in olive oil for about five minutes. Stuff the cleaned shells with meat. Seal the shell opening with butter. Place under the broiler for a few minutes, until the butter bubbles. Serve immediately. Serves 4.

FRIED SNAILS

2 cups snail meat, cleaned and blanched
1 cup water
Dash salt
2 bay leaves
½ tsp. parsley
½ tsp. thyme
2 cloves garlic, minced
Dash allspice
1 cup bread crumbs
Dash pepper
2 tsp. garlic powder
1/8–1/4 cup olive oil

Simmer snail meat for 10 minutes in water seasoned with salt, bay leaf, parsley, thyme, garlic and allspice. Roll the cooked snail meat in fine bread crumbs seasoned with salt, pepper and garlic powder. Fry in oil until browned. Sprinkle with lemon juice before serving.

SNAILS IN CREAM SAUCE

2 cups shelled, cleaned and blanched snails
1½ cups water
Thyme, salt, parsley, garlic and onion to taste
⅓ cup melted butter
2 Tbsp. minced green onion
3 Tbsp. whole wheat flour
2 cups milk (or one cup milk and one cup water from cooked snail meat)

Simmer snail meat for 10 minutes in the seasoned water. Drain and retain 1 cup of the liquid for cream sauce, if desired. Sauté minced onion in butter in a saucepan. Add flour and mix into a smooth paste. Blend in the milk or combination milk and seasoned water. Stirring constantly, cook over low heat until the sauce is thick. Add the snail meat and season with salt and pepper. Continue cooking until the snail meat is heated. Serve over toast or rice and garnish with chopped parsley. Serves five.

INSECTS

Mental barriers that keep us from using insects for food are as great or greater than the barriers over worms and snails, and are not easily broken. Aversion to eating insects is strictly acquired; anything acquired can be dropped.

The protein content of insects far surpasses that of our conventional meats. Termites are 40% protein and grasshoppers are 60%, compared to the 20% of either chicken or beef. Insects have surprisingly good tastes. Mashed and added to soups, stews, rice dishes or baked into bread, their taste is hardly noticeable. Prejudices based on insects being unclean, unpalatable or nonnutritious are entirely without basis.

A large grasshopper roasted on a stick was a delicacy to the early Southern California Indians. Even today, as you travel through Mexico and other countries, you will see baskets full of fried grasshoppers in the marketplace, and children eating grasshoppers, just as you see children here munching on potato chips. Remove the wings, legs and head of the grasshopper before eating. Add fried grasshopper, finely chopped, to your favorite bread recipe for subtle flavor enhancement. In ancient Greece and Rome grasshoppers, locusts and cicadas fried in oil were considered a finer delicacy than the best meat or fish.

Large ants fried whole taste excellent. Try grinding some into a paste or powder and mix into other foods. You are likely to find the larvae of beetles and termites in rotting wood. Raw, roasted or boiled, they make good eating. Dragonflies fried with onions and shrimp are considered a delicacy in Japan and Malaysia. Eaten raw after the wings have been removed, dragonflies have an unobjectionable nutty taste.

Raw potato bugs (Jerusalem crickets) taste only slightly unpleasant, like a rotten green pepper; fried they taste quite acceptable. Since tomato worms (the larvae of the sphinx moth) eat the toxic leaves of tomatoes and Jimson weed, I boil the tomato worm, discard the water, and reboil with a bay leaf and a piece of watercress. The meat of the tomato worm tastes surprisingly like shrimp.

South American Indians eat raw centipedes over a foot long and consider this treat especially good for youngsters. Caterpillars are edible, but leave the hairy ones where you find them.

NATURAL INSECT CONTROL

Truly healthy soil is not just "dirt" into which we insert plants. Within the soil lives an entire life system that is far more complex than we normally realize. Much of what the gardener considers "bad" for the garden, may seem unaesthetic, but is not bad at all. An abundance of insects eating your tomato leaves may indicate poor soil in need of nutrients. Consider insects vital messengers for your soil.

Killing off insects means killing off the messengers—the soil remains the same—so you can expect more insects in the future.

Killing off insects is dealing with symptoms, not causes. Insects attack only the weakest plants; healthy soil has far less insect infestation. Improve your soil and you remove insects from your garden naturally and ecologically.

Grow sunflowers in your garden and let them stand as an open invitation to the birds. The birds will feed on your garden pests. If you have the space, you might consider installing a simple bird house, bird bath and bird feeder to encourage birds on your property. Don't kill nonpoisonous snakes or toads in your garden. They eat the insects you find undesirable.

Natural insect control is a feasible way to keep your unwanted insect population in check. Ladybugs, lacewings, preying mantids and trichogramma wasps are popular natural insect controllers. Many nursuries and mail order companies sell these insects.

Aromatic plants, such as marigolds, also repel insects. Tansy repels ants. Garlic plants are highly repellent. Other aromatic herbs to plant include rue, thyme, oregano, marjoram, lavender, and sage.

If you need to use a spray on any insects, make it from natural materials such as red pepper, onion, garlic, tobacco and tea. For something stronger (which really should not be necessary) use natural insecticides such as rotenone, pyrethrum and ryania. Put out little bowls of beer to attract and drown the snails (if you decide not to eat them). Dust cabbage with wood ashes and clean plant leaves with a spray of water for pest control.

Bugs are part of the ecosystem—you can't kill them all off—and all of them are not bad. See bugs as messengers telling you what you must do about your soil, because, in reality, the most important aspect of insect control is having good-quality, living, vibrant soil, full of worms and other microorganisms.

SUGGESTED READING

DeBach, Paul, ed. *Biological Control of Insect Pests and Weeds.* New York: Reinhold, 1964.

Gaddie, Ronald E. *Earthworms for Ecology and Profit: Earthworms and the Ecology.* Bookworm Publications, 1977.

Haynes, Hank. *There's a Fortune in Worms.* Los Angeles: Brooke House, 1976. This small 60 page book offers the home worm farmer everything he or she needs to know to begin raising worms for gardening, food or profit. Haynes tells the reader how to market the worms also.

Lauber, Patricia. *Earthworms: Underground Farmers.* Champaign, IL: Garrard Publishing Company, 1976.

Minnich, Jerry. *The Earthworm Book.* Emmaus, PA: Rodale Press, 1977. How to raise and use earthworms for your farm and garden.

Shields, Robert F. *Earthworm Buyer's Guide.* Shields, IL: Shields Publishing, yearly editions, 1976.

III
WILD CITY PLANTS

One should taste wild foods often and do it with a positive attitude. Many people die amid plenty simply because they cannot "stomach" wild foods. The stomach only reacts to the stimulus given by a prejudiced brain in these cases. A few people often develop mental defense mechanisms against eating [wild food] . . . Needless to say, these defenses can kill even a healthy person lost in the wilds.

Larry Dean Olsen,
Wilderness survival expert

I doubt if there is anyone who hasn't heard the story of the child who, when asked where milk comes from, answered "from the supermarket." We grownups chuckle at this delightful naiveté; but because most urban dwellers are isolated from food production, the response may not be as childish as it sounds. How many adults know where ingredients come from? How many could feed themselves, if a major disaster cut off their supermarket source? Most of us would be no better off than the naive child.

How does one begin plant study? Learn the plants, weeds and trees that grow in your neighborhood, lawn, garden and local vacant lots. Throughout the United States there is an abundance of edible plants close at hand, just waiting to be harvested. Previous knowledge of wild foods is invaluable, if you find yourself in a survival situation.

Nature provides us with an abundance of free food in the form of "weeds." Weeds are those plants we object to, those plants we do not want around. But wild, medicinal and healthful plants are not things of the country and forest only; they are here, in the city, free for the picking. Let's reevaluate these plants in light of their potential uses.

Vacant lots, untended gardens, roadsides, yards and parks are all excellent places to gather wild food. These plants need only be identified accurately, gathered and properly prepared. They will liven up your meals with new tastes, packed with vitamins and minerals, and are fresher than anything you can buy in the supermarket or health food store. Soups, teas, stews and many other foods can be made from these healthful and free plants, found everywhere. Picking them provides a sense of self-sufficiency that one can usually only get by working hard in a garden. Wild plants can also be used as vegetable dyes, beauty aids and for decoration.

Take your children with you. Learn about these plants together; teach them the ways of nature, even in the city. Don't wait to get "back to the country" to learn about nature. Start now. Take a walk and discover where you can locate wild food in your area. You will be surprised at the abundance of food all around you. You will become less dependent upon the supermarket, and may even prefer this kind of food. It is fresher, more alive than food which has been picked and trucked, handled and rehandled, finally sitting wilting, waiting for you to buy it and take it home.

ASSAULT ON THE WEED

As the early American settlers began their westward move across North America, they brought (intentionally and unintentionally) a large array of European plants. Many of these plants preferred the disturbed and cleared soil that resulted when a settlement was established. One particular plant was called "White Man's Foot" by the Indians, not because it resembled a foot, but because it appeared wherever the white man went. It preferred environments that had been cleared, leveled and reclaimed from the wilderness. Truly, this was a plant of civilization.

White Man's Foot or Plantain is still with us today, as common as the well known dandelion. Plantain and a broad assortment of other plants have come to prefer the cities. They are maligned as weeds, nuisances and pests.

Weeds were brought to North America because they were medicinal as well as edible. Our ancestors would have yelled in outrage, if they saw the vast array of techniques and herbicides their progeny has devised to kill these wonderful lifesaving plants. "Our descendants," they might say, "know not what they do. They desecrate our very memories by destroying the plants we so lovingly nurtured for them."

"Establishment agronomists" are united in an all-out assault on the weed. When Euell Gibbons was still on the scene, he made the public aware that there are literally millions of tons of wild food fighting to stay alive in what appears to be a losing struggle. Gibbons said the main reason people do not use wild food is fear of ridicule; "stooping" to pick what everyone else treats as trash. Gibbons managed to broadcast his message as part of a television commercial for several months in 1975. The commercial was halted on Independence Day, 1975, when the Federal Trade Commission asked General Foods Corporation to withdraw its Post Grape Nuts ads in which Gibbons described certain plants as edible. In one ad Gibbons grabbed a pine branch saying, "I've spent years learning about natural foods. Ever eat a pine tree? Many parts are edible. Natural ingredients are important to me. That's why Post Grape Nuts is part of my breakfast."

The Federal Trade Commission felt that four of the ads had "the tendency or capacity to influence children to eat plants or parts thereof which they find growing or in natural surroundings. Some of the plants or parts thereof are harmful if eaten."

When the FTC announced their veto of the ads, they said that the ads "undercut a commonly recognized safety principle, namely that children should not eat any plants found growing in natural surroundings, except under adult supervision."

Although Gibbons broadened acceptance of these weed foods, the commercials also prompted a series of jokes and satires on the edibility of park benches and highways. Gibbons represented the weeds of America. What we were saying about Gibbons we were saying about weeds—or so it seemed. Gibbons often said he did not mind the jokes; he laughed all the way to the bank (he was making many lecture and television appearances and his popular books were selling a combined total of 100,000 copies a year). Almost six months after the FTC ruling on his commercials, Gibbons died at age 64. As expected, it began again. "Euell Gibbons died of wild food, of weeds," everyone was saying. "If I eat wild foods, I'll die too."

With all respect to the good intent of the FTC ruling, it succeeded in generating a wave of mistrust and fear of wild foods in general, even natural foods. Gibbons had stressed that one must *never* eat plants (or parts of plants) until they are positively identified as edible. Unfortunately the media followed the FTC ruling with headlines of two cases in the vicinity of the Angeles National Forest in Los Angeles County where teenagers, glamorized by Gibbons "living off the land" ads, ate toxic weeds. The public had renewed distrust of weeds, Gibbons and wild, natural foods.

Since 1974, when I began teaching outdoor classes in the identification and use of wild plants for food and medicine, I have been confronted continually with this issue. Many people are interested in becoming self-sufficient with city skills and back-country skills. True self-sufficiency cannot be accomplished without the use of wild plants. You must accept the possible dangers, but with education and care you should have no problems.

Almost all the weeds in your garden, nearby empty lots and along roadsides are good food and acceptable emergency medicine. European weeds have naturalized themselves so well that they now come up in sidewalk cracks, through blacktop, in sprayed and cultivated gardens, in carefully tended containers next to the imported shrubs, in freeway and parkway strips—wherever a sliver of an opening and a little moisture appear.

Dandelion, lamb's quarters, pigweed, mallow, mustard, dock, and sow thistle (all excellent edibles from Europe) are some of the most common "intruders." Those people who are professionally or personally preoccupied with aesthetics view this prolific growth as an altogether unhappy prospect. These weeds have taken over and seem to defy eradication.

One of the most persistent wild European weeds is chickweed. Even to the most pampered palate chickweed tastes delicious in salad. You can understand my dismay when, on the front page of "Weedone Illustrated News," distributed to all the major newspapers in the Spring of 1976, was the headline "Weedone Chickweed Control." American scientific genius had perfected a liquid, one pint of which mixed into ten gallons of water can kill all the chickweed on 3,000 square feet of lawn. Why our continued myopia? On the back page of this antiweed newspaper was a large chart with illustrations of 35 common weeds, nine herbicides and the most effective poison to kill each weed. Included in this list of "pest plants" was wild garlic, chufa, chickweed, plantain, sorrel, dandelion, yarrow, sourgrass and ground ivy.

Why is America killing off its herbal birthright? With the torrent of research dollars poured into eradication, I fear that one day soon big business will succeed. Although some plants take an awful lot of poison to kill, if we are so determined to kill off these volunteer plants, no doubt we soon will, toxifying our soils in the process.

I see beauty in weeds. I know they are valuable. My yard will never be manicured and hacked and pummeled into a "picture of beauty." I allow weeds to grow because I see their virile independence, the courage to persist boldly in the face of public censure. In exchange for the safe haven I provide them, they gladly offer their tithes of an occasional leaf, flower, bud or stem for my sustenance, plus 12 to 14 hours a day of the most delicious oxygen and cooling vapor mist.

Rather than looking to real solutions, such as plant identification classes in elementary schools, our minds go to Hitler's "ultimate solution" level of thinking. How shocked these screamers for wholesale plant slaughter would be, learning that their Christmas holly, mistletoe and beautiful poinsettias are also poisonous. Would they have asked for banishment of Easter celebrations because the Calla lily is poisonous, and for eradication of formal gardens because some hybrid ornamentals are poisonous as well?

The battle rages on; some of us fight, some are passive, and others accept the voice of whoever speaks loudest and longest. Wild plants will take over any yard if the owner doesn't work to keep them away. Take a month's vacation and their "scouts" will be everywhere. Allowed to do their natural work, they would cover the earth green and provide food for us all. The earth has been their home far longer than it has been ours.

SUGGESTED READING

Gibbons, Euell. *Beachcomber's Handbook.* New York: David McKay Company, 1967.

———. *Stalking The Blue Eyed Scallop.* New York: David McKay Company, 1964.

———. *Stalking the Healthful Herbs.* New York: David McKay Company, 1966.

———. *Stalking the Wild Asparagus.* New York: David McKay Company, 1962. For readability and variety of information, Euell Gibbons' books are a must. The books cover identification, use, and Euell Gibbons' personal and valuable insights. Subjects are wild foods, herbs, sea food and food of the Pacific Islands.

Harris, Ben Charles. *Eat the Weeds.* New Caanan, CT: Keats Publishing Company, 1961. A great little book with valuable information on city weeds. Covers 150 common plants.

A WILD THANKSGIVING

On Thanksgiving 1976 I took part in a very special dinner. A group of about thirty people identified, gathered and prepared wild food for a Thanksgiving wild food dinner.

We had a huge salad of dock leaves, cattail shoots, Jerusalem artichokes, watercress, wild radish leaves, purslane, raw prickly pear pads, fennel stalks, sorrel, nasturtium and wild celery. This incredible mixture was lightly seasoned with onions, oil and vinegar. Watercress soup was my favorite food of the day. Acorns, that had been leached and dried, were ground up in a hand grinder, and added to carob and wheat flour for delicious loaves of acorn bread. The acorn bread disappeared quickly!

We peeled and diced prickly pears, mixed them with onions, and fried the mixture until all the juice was gone. After adding about two dozen eggs, we had a prickly pear omelette. We breaded and deep fried sulphur mushrooms. These colorful orange mushrooms were chewy and tasty.

A large pot of Jerusalem artichokes (edible tubers of a sunflower species) from my garden were boiled and eaten like potatoes. A large bowl of sweet carob pods and two bowls of quartered pineapple guavas sat on the table for all to enjoy. Outside on a smaller stove someone prepared a mixed stew-fry of cattail shoots, purslane and various other greens. Seasoned with onions and tamari sauce, it was a popular dish with everyone. Mormon tea and yerba santa tea were the hot beverages. People brought various breads from home, such as sage bread, persimmon bread, and pumpkin bread. Someone even brought some homemade horehound syrup.

The group was enthusiastically involved in their Thanksgiving alternative. After the meal, we gathered together and talked. A member of the American Indian Movement talked to us about the proper ways of gathering plants. He spoke of the inseparable spiritual awareness that goes hand in hand with learning about and using wild plants; how plants should be gathered thoughtfully, intimately thanking the plant, leaving something for the plant (such as tobacco) in exchange. He encouraged the group to continue living a natural and harmonious lifestyle that serves to protect and not destroy the earth.

To many in the group gathering and preparing their own meal from the wild was a new experience. A few had been skeptical of what kind of meal we could actually prepare.

Because everyone became personally involved in gathering and preparing the meal, everyone felt an integral part of this Thanksgiving ceremony. The group welcomed many more such meals to increase feelings of self-sufficiency. A sense of freedom filled the air—freedom from the urban wilderness while still living in it. Individual thanks were directed to the Great Spirit which has provided us all with the plants needed on our path of spiritual evolution.

We discussed the concepts (and misconceptions) "primitive" and "civilized." Specific personal actions practiced by enough individuals can successfully challenge and defeat much of the ignorant and destructive mentality so rampant in the world today. We took with us the seeds that have the potential to cover the earth in a forest of sanity. I thought of the words from Kahlil Gibran's *Prophet:* "Who can separate his faith from his actions, or his belief from his occupations?"

GATHERING URBAN WEEDS

Make positive identification before gathering any plants. Know what you are going to do with the plant. *Make sure you know the plant or don't use it at all.* There is no quick rule for identification or for differentiating poisonous and edible plants. You will need to learn them individually, either through books, classes, or competent naturalists and botanists.

Whenever possible, pick the youngest and most tender leaves of a plant; older leaves tend to become tough and bitter. Your taste buds will prefer the young leaves, and you will like wild foods better.

When gathering leaves for salad or cooked greens, you need not uproot the entire plant. Pick the leaves carefully, so the plant will continue growing and go to seed. Avoid stripping the plant bare. Beauty can remain, if you gather with care. Think of the next foragers, the birds and animals, and the next generation of plants. Consider the relative abundance of the plant. If there is only one or a few, leave them. Abundance is no reason to pick and grab carelessly. Gather only what you need or will use.

While digging tubers and taproots, remember that the plant will no longer exist. Put small roots and tubers back into the soil, so they may continue growing. Show concern for these living things. Forage in your garden, vacant lots, canyons and fields.

The following is a compilation of the most common and most available free city foods. Use the plant identification chart for any vocabulary or descriptions you are not familiar with. If you have any question about wild plant identification, please double check with another reference.

BURDOCK *(Arctium lappa)*

Burdock is a biennial plant which grows commonly in the Eastern United States. It is commonly found along fence walls, roadways,

Burdock

waste places, vacant lots and generally in populated areas, especially near human dwellings.

The domesticated burdock plant is called *gobo*. The root is a common ingredient in Sukiyaki. Gather the roots from the first year nonflowering plant only. Peel the roots of the outer rind and then slice them. Cooked till tender, they can be added to soups, stews or eaten alone.

CHICKWEED *(Stellaria media)*

One of the most common weeds in the city after heavy rains is chickweed. Chickweed is usually available from midwinter to late summer. It is a native European plant that has naturalized throughout the entire United States, Central and South America. Chickweed usually grows in moist, shady places. Look for it under trees, on the north side of your house, and growing under other plants in your garden.

This annual has a weak stem up to a foot in length with about 1/16 inch diameter. Under close observation, you can see a line of tiny hairs running along one side of the stem which changes to the opposite side of the stem at each node. Each chickweed leaf is oval in shape, with an untoothed margin, the tip coming to a sharp point. The leaves grow opposite each other. The leaves on the upper part of the plant lack a petiole (or stalk), whereas the lower leaves possess a petiole. The positive identifying feature is chickweed's tiny, white flowers which are about ¼ inch in diameter. At first glance it appears that there are ten petals; closer observation reveals only five petals, each with a deep cleft in the middle.

To gather, use a knife. Cut a few handfuls from a thick patch, being careful not to pull out the roots. Carefully pick out any foreign matter that you might pick along with the chickweed. After a brief cold water rinse, this succulent salad plant is ready to eat.

Although chickweed is acceptable cooked like spinach, it is best eaten raw in salads. The entire above-ground plant can be eaten. Chickweed has a fresh and mild taste which is similar to lettuce and rarely objectionable. You can mix this vitamin-C-rich plant into your regular salad, or you can try a salad made entirely of chickweed.

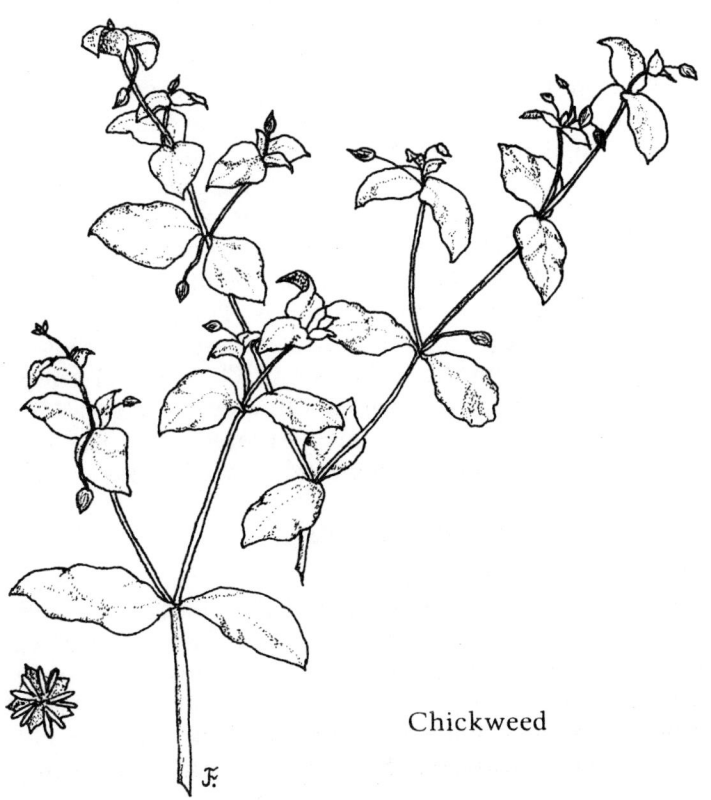

Chickweed

CHICKWEED SALAD

Large bowl fresh chickweed torn into bite-size pieces
2 cloves garlic, diced
1 avocado, cut in small pieces
3 cherry tomatoes
2 or 3 sprigs parsley or watercress, finely chopped

Mix into a bowl and add prepared dressing.

LIGHT DILL DRESSING

3 Tbsp. wine vinegar
3 Tbsp. oil
1 tsp. powdered dill weed
Dash pepper
Dash paprika
Dash sea salt

Mix all ingredients together and add to the Chickweed Salad.

Two plants you may confuse with chickweed are scarlet pimpernel and common spurge. Scarlet pimpernel or red chickweed *(Anagallis arvensis)*, when not in flower, looks deceptively like chickweed, except that it is less delicate. Unlike chickweed, all of scarlet pimpernel's leaves are sessile, meaning they lack a stalk and are attached directly to the stem. It has a deep orange-red, five-petalled flower which opens only in sunny weather. Raw scarlet pimpernel tastes bitter, but some people use it as a cooked green. The plant is suspected of mildly poisoning horses and livestock that graze on it. The human effects of eating this plant include diarrhea and nausea. Scarlet pimpernel has medicinal properties, but it is *not* a salad plant.

Common spurge *(Euphorbia peplus)* also looks similar to chickweed and often grows right beside it, but there are several distinct differences. The main stem of common spurge is nearly erect, and the leaves lack the sharp-pointed tip of chickweed. The flowers of common spurge are very inconspicuous. For positive identification, break the stem of the plant you suspect is spurge. If it is spurge, a thick white sap will drip from the broken stem. The sap of all the *Euphorbias* can cause skin irritation and partial blindness, if it gets into the eyes. The plant has toxic effects, if eaten in large amounts. These effects include nausea, vomiting, diarrhea and stomach pains.

CHICORY *(Cichorium intybus)*

Chicory (also known as succory or endive) is a perennial weed that grows throughout the United States and abundantly in urban areas. You will find chicory in vacant lots, along roadsides and waste areas, in dry soil or gravel. Chicory will grow in any soil and is even a good kitchen garden plant.

Chicory has long lanceolate, sparse, dark gray-green leaves that are coarsely-toothed. It branches abruptly, has smooth upper stems,

Chickory

bristly lower stems, and has a stiff, angular appearance. The plant reaches one to three feet high and has a long tap root. Chicory has many beautiful voilet-blue flowerheads which bloom continuously in the Northeast, June through September. The flowers open and face the rising sun, wither by the afternoon, and send out new flowers by the following morning.

    Roasted and ground chicory root beverage is common as a coffee substitute and mixed with coffee, especially in the hot and drowsy, humid South where people drink large amounts of coffee. Coffee alone can be disasterous to one's body, particularly to the liver. Chicory, on the other hand, is an excellent plant for the liver. Chicory was added to coffee originally to counter coffee's negative effects. One-half to three ounces of ground chicory root added to a pound of coffee will add an extra ten cups of coffee, while enhancing the flavor.

To prepare chicory for beverages, wash the roots with a brush. Slice the roots in thin strips; tie the strips together with wire or leave them untied to dry in an oven, an attic or in direct sunlight. Roast the dried roots for about an hour in an oven or put tied roots over an open fire. Grind the roasted roots in a blender, coffee grinder or mill.

Chicory leaves are great as a raw salad vegetable, steamed alone or together with other vegetables, sautéed, deep fried in batter, and added to soufflés or soups.

PICKLED CHICORY

2 cups chicory buds or flowers
3 cups apple cider vinegar
½ tsp. sea salt
¼ tsp. ground cloves
½ cup honey

Wash the flowers in ice cold water, and place them in a sterilized canning jar. Combine salt, cloves and vinegar, and bring to a boil over medium heat in a nonaluminum pan. Simmer for 5 minutes. Remove from heat; add honey and stir until dissolved. Pour over flowers. Seal the jar and store in a cool dark place for at least one week.

CLOVER, RED *(Trifolium pratense)*

Red clover, also known as bee bread, is a perennial of short duration which grows throughout the United States in urban locations: in vacant lots, parks and along roadsides. Clover prefers light, sandy areas but will grow anywhere, five to eight inches tall.

Collect clover greens in the spring when the blossoms are full, and add them to your mixed green salads. Cook them with other greens.

Red clover flowers range in color from pink to brown. Indians used to grind up the flowers and use the meal as flour. The dried blossoms make a delicious tea, hot or cold.

DANDELION *(Taraxacum officinale)*

Dandelion or wild endive is a perennial of the chicory family which grows heartily almost everywhere across the country, in yards, gardens, vacant lots and along roadsides. Dandelion has shining green leaves, a long narrow stalk and a flower stem which is longer than the leaves. Dandelion bears a single yellow flower, April through November. To avoid bitterness pick the leaves before the plant flowers. The roots can be picked throughout the year.

Eat the healthful greens raw, in salads, steamed with other vegetables, or cooked into soups and stews. Dig up the long, deep, milky tap root and use it, either alone or mixed with other grains, for a hearty and popular noncaffeine beverage.

Wash the roots thoroughly with a brush; dry and roast as you

Gathering Urban Weeds

Dandelion

would other roots, in the oven at a moderate temperature. Do not roast too fast. Grind in a blender, coffee grinder or hand mill. Percolate, drip or boil and strain; prepare as you would coffee or other grain beverages.

Children love to blow the delicate flower at each other and call it a "blow ball." The flower provides a natural yellow dye, and food for bees. The flower opens early and closes at dusk; it reacts sensitively to weather, closing during rain.

DANDELION WINE

- 2 qts. fresh dandelion petals
- 2 qts. boiling water
- 1 large orange, thinly sliced
- 2 medium lemons, thinly sliced
- 1 lb. honey
- piece of whole ginger root
- ½ tsp. brewer's yeast

Wash and dry the petals and put them into boiling water. (Do not use an aluminum pan.) Cover with a clean cloth and let sit for 3 days, stirring frequently. Strain into a second clean pan and add the orange and lemons. Add honey and ginger; boil for half an hour. Return to first pan and let cool. When lukewarm, spread the yeast on a piece of toast, and add it to the wine. Let sit for 2 days. Pour into a crock and keep covered for 2 months. Bottle and let sit for another 2 months in a cool, dark place, turning the bottles occasionally.

FENNEL *(Foeniculum vulgare)*

An important wild salad plant in the Western states is fennel. This ferny member of the parsley family is most commonly found along the Pacific coast in canyons, streams and vacant lots, but it is scattered

Fennel

throughout the West in a broad variety of localities. Some members of the parsley family are deadly poisonous; extreme care must be taken in identifying this plant accurately. Identifying features are fennel's strong licorice smell and blue-green color.

Fennel is a perennial which in midwinter begins producing a ferny cluster of leaves, finely divided into linear segments. As this ferny cluster matures, the leaves get larger; by late spring the flower stalk reaches its top height of seven to eight feet.

Fennel's flower cluster is an umbel (typical of the parsley family), with more or less flat-topped clusters of flowers whose stalks all have a common origin. In spring the individual flowers are yellow, and by late summer the prominent seed is mature. The stems of this plant are striated, similar to a celery stalk. The base of each fennel leaf also resembles celery in the way it broadens and clasps the main stem.

The young fennel stalk offers one of the most deliciously unique tastes available in wild foods and is superb in salads. This tender section may be up to 1½ inches thick and has a sweet licorice taste. Dice and add to your favorite salad mixture. You can use fennel stalks for about ¼ to ⅓ of the salad's bulk. Fennel mixes well with virtually any leafy salad.

NOTE: Hemlock *(Conium maculatum)*, another member of the parsley family, is deadly poisonous. The pinnately-compound leaves of poison hemlock, with its lanceolate to oblong leaflets and serrated margins, in no way resemble the linear segment of fennel leaves. Hemlock does not have the characteristic licorice smell of fennel and has purple blotches or streaks on the stalks.

Hemlock—*Danger*

In spite of these clear distinctions, hemlock is occasionally mistaken for an edible plant, and eaten by campers who are either misinformed or uninformed.

The cardinal rule for wild food foragers is: *never eat any plant that has not been positively identified as an edible species.*

LAMB'S QUARTER *(Chenopodium album)*

Lamb's quarter is another common garden and city weed, found throughout the United States, under hedges, in vacant lots and even in the cracks of sidewalks. Lamb's quarter is identified by its erect stem,

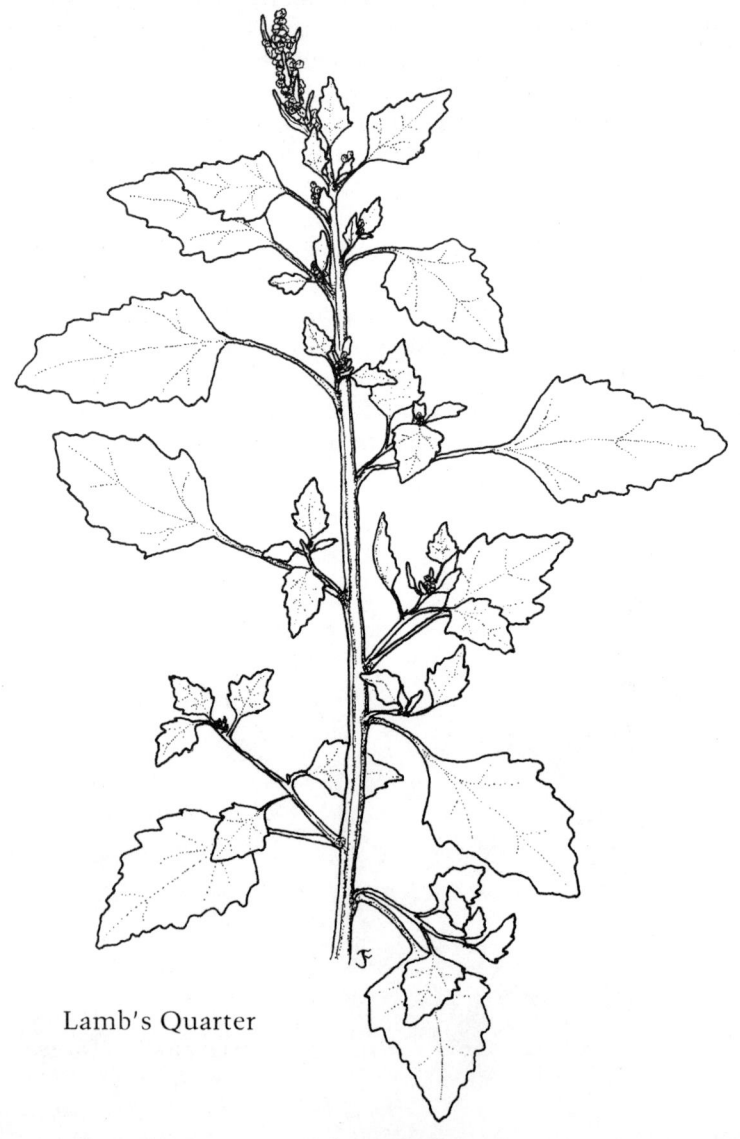

Lamb's Quarter

alternate ovoid to triangular-toothed leaves and a white mealiness that covers the entire plant. The plant is a relative of common beets and spinach and is rich in vitamin A, C, and calcium. Pick this plant when it is a few inches tall to three to four feet (early spring to late fall).

Lamb's quarter cooks like spinach, but tastes even better. Season with onions or butter. Like spinach, it will cook down to half its size. The raw leaves make a fresh, tasty and healthful addition to your wild salad.

MILKWEED *(Asclepias syriaca)*

Milkweed grows along roadsides, waste areas, vacant lots, and generally in open spaces in the city. Milkweed has ovate leaves (four to six inches long) and flowers from June through September. The plant grows to about three to four feet tall. The stems have a milky white sap which accounts for the plant's name.

Add the young shoots to boiling water, cook for a few minutes, drain, and reboil. Sometimes even a third boiling is required to eliminate bitterness. Young milkweed pods can be prepared in the same fashion. The unopened flowers should also be boiled, and they are good served with butter on macaroni.

Milkweed

NETTLE *(Urtica dioeca)*

Nettle grows wild in most of the United States in empty fields, near fresh water and especially near raspberry plants. Identify nettle easily by its sting. Brushing against nettle breaks the tips of the small needles covering the plant and releases an organic liquid irritant. The formic acid of the liquid causes an immediate burning sensation, a subsequent stinging and a welting rash, but within two to four hours all effects are gone.

Nettle is rich in vitamin A and C and is an exceptionally rich source of protein. Only the first nettle greens of spring should be eaten, carefully gathered with a knife and gloves. Older plants become less palatable. Cooking the tender tops (steamed or boiled) completely removes the nettles' stinging properties. Nettle is quite delicious served simply with butter. Dried nettle leaves make an excellent tea which is especially good for the hair and skin. The dried plant is excellent food for goats, chickens, horses and livestock. Since nettle

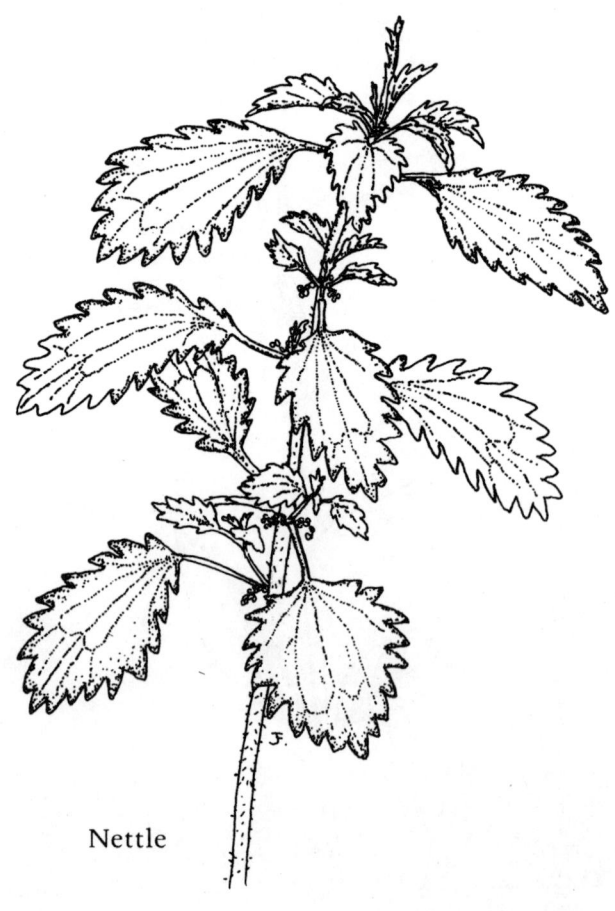

Nettle

contains approximately seven percent nitrogen (based on dry weight), it is an ideal plant to use as a mulch or fertilizer for your garden. The presence of nettle indicates rich soil.

Their stinging aside, nettles are one of the most nutritious foods in the plant kingdom and provide a source of quality fiber. English poet Thomas Campbell has claimed: "In Scotland, I have eaten nettle, I have slept in nettle sheets, I have dined off a nettle tablecloth."

PLANTAIN *(Plantago major, Plantago lanceolata)*

Plantain is a hearty plant that will survive where others cannot. It thrives in vacant lots, lawns, along streets and roadsides and is difficult to uproot. The Indians called this plant the "White Man's Foot," because it preferred the disturbed soil of the new settlements and spread westward with migration.

Plantain is a low-growing plant whose dark green leaves grow in a rosette. The leaves are vertically ribbed with many veins. From the middle of the plant shoots a flower spike. Plantain blooms from June to November in the Eastern United States and longer in more temperate regions. Eat it raw in salads or cooked like spinach. Plantain has a reputation as a medicinal plant, especially as a wound herb when applied externally.

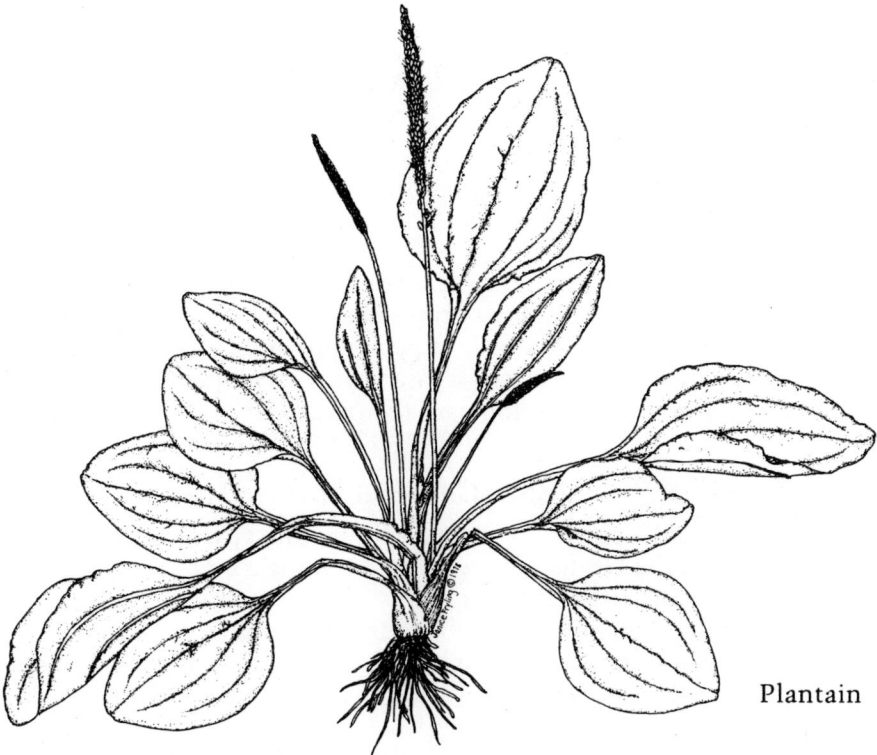

Plantain

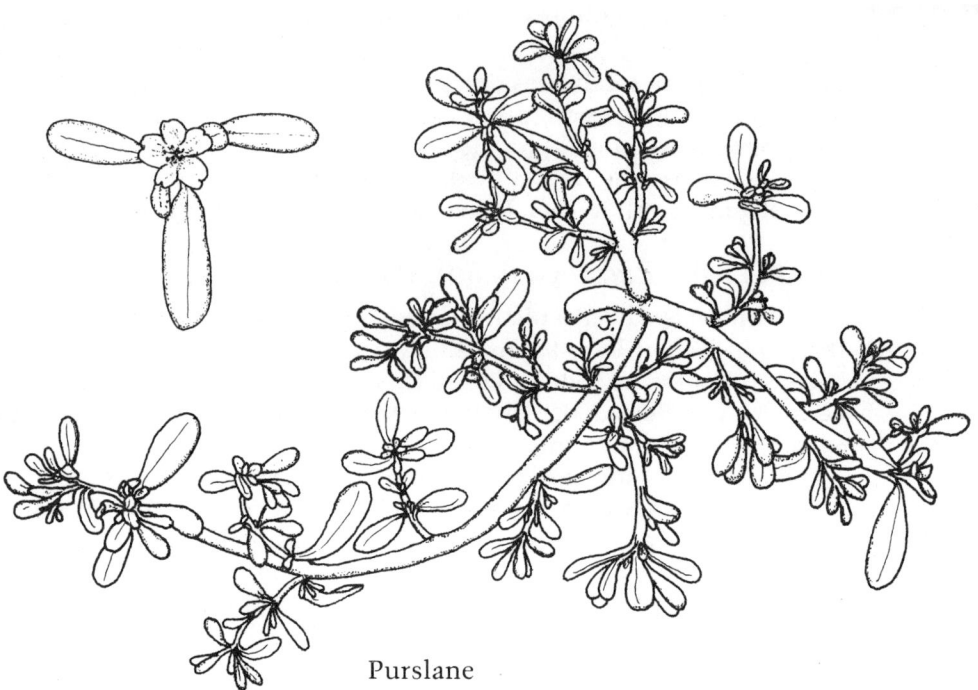

Purslane

PURSLANE *(Portulacca oleracea)*

Purslane is a succulent that grows prostrate with red branching stems and paddle-shaped leaves. It grows throughout the United States. Eat purslane raw (slightly slimy), fried, boiled, or added to soups, stews and omelettes. The large purslane stems can also be eaten.

Thoreau once wrote, "I learned that a man may use as simple a diet as the animals, and yet retain health and strength. I have made a satisfactory dinner off a dish of purslane which I gathered and boiled. Yet men have come to such a pass that they frequently starve, not for want of necessaries, but for want of luxuries."

SHEPHERD'S PURSE *(Capsella bursa-pastoris)*

Shepherd's purse is a member of the mustard family which grows throughout the United States in fields, along roadsides, in ditches and upon heaps of earth. It has a sturdy taproot, four-petalled white blossoms and propagates March through November, when the plant is blooming and producing fruit. Most conspicuous is the plant's heart-shaped seed pods. The mild-tasting leaves are excellent in salads or lightly steamed.

SORREL, SHEEP *(Rumex acetosella)*

Sorrel grows everywhere in the cities, along roadsides, grassy areas and in vacant lots. Its presence indicates poor soil. This plant has slender flower stems, small flowers, dark green leaves in a rosette, and will grow six to twelve inches tall. Sorrel has shallow but extensive roots and is a perennial.

Sorrel is a thriving and surviving plant that will literally grow anywhere, which is why its presence as a weed is so plentiful. Dig up the plant with its roots and some of its soil; take it home and grow it in a flower pot or window box. Water sorrel often and give it acid soil. Sorrel is propagated by root division.

Use sorrel as you would spinach; it is slightly acidic, but very good-tasting and loaded with vitamins and minerals.

Sheep Sorrel

SORREL SOUP

1 Tbsp. oil
1 medium onion, chopped
1 carrot, sliced
1 mushroom, sliced
2 cups water or stock
2 cups chopped spinach
1 cup chopped sorrel
2 Tbsp. fresh sweet basil
(1 Tbsp. dried)
1 Tbsp. miso or tamari dissolved
  in 1 cup water
¼ tsp. salt

Sauté onion, carrot and mushroom in oil. Add water or stock. Bring to a boil, reduce heat, cover, and simmer for about 20 minutes or until the carrot is tender. Add chopped sorrel, spinach and sweet basil. Cover again and simmer for about 3 minutes.

Dissolve the miso in a cup of warm water. Add the miso and water mixture to the soup. Heat just enough to warm the soup. Add salt to taste. Makes 3 cups.

SORREL SOUFFLÉ

¼ cup cornstarch
¼ cup butter
4 egg yolks
4 egg whites
1 lb. sorrel, steamed, drained, chopped
¼ cup chopped onion
⅛ tsp. nutmeg
¼ tsp. salt
¼ tsp. cream of tartar
3 Tbsp. Parmesan cheese

Preheat oven to 400°. Prepare sauce, mixing cornstarch and butter in a saucepan over moderate flame. Beating constantly, gradually add egg yolks. Stir in sorrel, onion and nutmeg.

In a separate bowl beat egg whites, salt and cream of tartar, until stiff. Fold in sorrel mixture. Pour into a greased 5 cup soufflé dish. Sprinkle Parmesan cheese over the top. Place on lower rack of oven. Reduce heat to 375°. Bake 50 minutes until set. Serves 4.

SORREL, WOOD *(Oxalis acetosella)*

Wood sorrel is also known as sourgrass or oxalis, and sometimes referred to as shamrock or four-leaf clover, although it is not a true clover. Wood sorrel is common throughout the United States. Children enjoy chewing on the sour-tasting succulent stems, although eating too much raw will cause stomach aches and possibly vomiting. Several species are found throughout North America with pink, white or yellow five-petalled flowers. Each leaf is composed of three (and sometimes four) heart-shaped leaflets.

This plant is delicious added to salads and imparts a vinegar taste. California Indians fermented the leaves of this plant to make a type of wild sauerkraut.

Wood Sorrel

PLANT USES AND PREPARATION

PLANT	PART USED	PREPARATION	USES
Alyssum, sweet *(Alyssum maritimum)*	flowers	raw	salads, garnish
	young leaves	raw	salads
		steamed, sautéed	vegetable dishes
		boiled	soups, stews
Artichoke, Jerusalem *(Helianthus tuberosus)*	whole tuber (root)	raw	salads
		boiled	soups, (mashed like potatoes), sauces
		sliced, fried	vegetable dishes
Asparagus *(Asparagus officinalis)*	new shoots	raw	salads
		steamed, sautéed	vegetable dishes, omelettes, good with cheese, creamed or uncreamed, soups

PLANT	PART USED	PREPARATION	USES
Burdock (*Arctium lappa*)	root ("gobo")	boiled	soups, stews
	young stalk	raw	salads
		boiled	soups, vegetable dishes
Cactus, Prickly-Pear (*Opuntia sp.*)	pad (nopale)	raw (peeled)	salads, drinks, candy, ice cream
		boiled	omelettes, casseroles
		fried, sautéed	omelettes
	seeds	dried	flour
	fruit (pear)	boiled	jams, pies,
		raw	drinks, ice cream
Carob (*Ceratonia siliqua*)	pod (fruit)	raw dried, roasted, ground into powder or flour	sweet snack food blender drinks, hot cocoa substitute, cookies, breads. (Use in place of sugar, chocolate or flour.)
Cattail (*Typha latifolia*)	starchy base of shoots	peel off fibrous covering and use raw boiled	salads (like potatoes)
	roots (rhizome)	dried, ground	flour
	stalk (inner portion), young white shoots	raw boiled, steamed	salads soups, vegetable dishes
	green flower spikes	boiled, steamed	(like corn-on-the-cob)
	pollen (top of spike)	shake into bag, sift	flour
Celery (*Apium graveolens*)	young leaves and stems	raw sautéed, steamed boiled	salads vegetable dishes creamed or uncreamed in soups
Chickweed (*Stellaria media*)	entire above-ground plant	raw boiled, steamed, sautéed	salads mixed with stronger greens, soups
Chicory (*Cichorium intybus*)	leaves	raw steamed, sautéed	salads alone or in mixed vegetable dishes
	stem	boiled, puréed dipped in batter & deep fried	soufflés tempura
	roots	roasted, ground	grain beverage (coffee substitute)
Clover, Red (*Trifolium pratense*)	blossoms	dried, infusion ground	tea flour
	young leaves	raw	salads

PLANT	PART USED	PREPARATION	USES
Dandelion (Taraxacum officinale)	young leaves	raw	salads, fruit salads, sandwiches
		boiled, steamed	mixed vegetable dishes, soups, fish dishes
	crown (upper root, lower leaf section)	boiled, steamed, baked	vegetable dishes with or without creamed sauce
	roots	roasted, ground	grain beverage (coffee substitute)
		boiled, steamed	vegetable dishes
	flower	fermentation	wine
Fennel (Foeniculum vulgare)	seeds	whole	tea, sprouts
		ground	licoricelike flavoring; breads, cookies, Italian soups, fish dishes, marinades
	tender base of stem	raw, diced	salads, (use like celery)
		boiled	stews, soups
Glasswort (Salicornia sp.)	young stems, upper sections	raw	salty taste to salad
		pickled	appetizer
		steamed	mixed vegetable dishes
Kelp	stems, air bladders	pickled	appetizer
	fronds	chopped, boiled	soups, sauces
		sautéed	mixed vegetable dishes
		dried & powdered	seasoning/salt replacer
Lamb's Quarter (Chenopodium album)	young leaves & stems	raw	salads
		steamed, sautéed	mixed greens, vegetable dishes, soups, stews, omelettes
	seeds	dried	flour
Lettuce, Miner's (Montia perfoliata)	entire above-ground plant	raw	salads, egg salad, sandwiches
		steamed lightly	greens
		sautéed	vegetable dishes
Lettuce, Sea (Ulva lactuca)	entire frond	raw	salads
		dried, whole & powdered	seasoning
		baked	stews
		boiled	soups, greens
Lily, Day (Hemerocallis fulva)	unopened flower buds	steamed, sautéed	vegetable dishes
	mature flower	dipped in batter, fried	appetizer

PLANT	PART USED	PREPARATION	USES
Lily, Day *(Hemerocallis fulva)*	flower, dried on plant	as is	seasoning, soups, stews
	underground tubers	raw, chilled boiled	salads (potatolike taste) soups, stews
Milkweed *(Asclepias syriaca)*	flowers	boiled	vegetable dishes
	young leaves	boiled	vegetable dishes, greens
	immature seed pod	boiled	vegetable dishes
Nasturtium *(Tropaeoleum sp.)*	flower	raw	salads
	leaves and stems	raw	salads
		steamed, sautéed	greens, vegetable dishes, omelettes
		baked	quiche
		boiled	soups, stews, jelly & jam
	seeds	pickled	appetizer (like capers)
Nettle *(Urtica dioeca)*	leaves	boiled	soups, stews, greens
Oak Tree *(Quercus sp.)*	Acorn	shelled, leached & osterized or ground	breads, cereals, flour
Orach *(Atriplex sp.)*	leaves	raw	salads
		steamed, boiled	cooked greens, soups
Plantain *(Plantago major and lanceolata)*	young leaves	raw	salads
		steamed, sautéed	vegetable dishes, greens
		boiled	soups, meat dishes
	seeds	soak, boil	(like rice)
Purslane *(Portulacca oleracea)*	seeds	ground	flour
	young leaves, stems	raw	salads
		pickled	appetizer
		sautéed, fried	omelettes, vegetable dishes
		boiled	soups, stews
Rose *(Rosa sp.)*	flower petals	raw	salads, a petal in wine, minced in pancake batter, pudding, muffins, mixed into omelettes
		boiled	jam and jelly
	fruit (hips)	boiled (fresh and dried)	tea
		raw	(like an apple)

PLANT	PART USED	PREPARATION	USES
Sea Rocket (*Cakile edentula*)	sprouts	raw	(mild horseradish taste) condiment, salads
		boiled	soups
	leaves	raw or creamed	(like horseradish)
Shepherd's Purse (*Capsella bursa-pastoris*)	young leaves	raw	(peppery seasoning) salads, sandwiches
		steamed, sautéed	vegetable dishes, omelettes
		boiled	soups, stews
	seeds	dried, whole	garnish
		ground	flour
Sorrel, Sheep (*Rumex acetosella*)	leaves	raw	salads
		steamed, sautéed	cabbage dishes, vegetable dishes, soufflés
		boiled	soups
Sorrel, Wood (*Oxalis acetosella*)	entire plant	raw	salads
		pickled	appetizer
Spinach, New Zealand (*Tetragonia expansa*)	leaves	raw	salads
		steamed	added to cooked seafood dishes
Violet (*Viola sp.*)	leaves	raw	salads
		steamed	greens, mixed vegetable dishes, stews
	flowers	boiled	jam, jelly
Watercress (*Nasturtium officinale*)	whole plant	raw	salads, garnish, sandwiches
		steamed, sautéed	vegetable dishes, shrimp dishes
		boiled	soup
		puréed	soups, sauces
		dried	seasoning

SUGGESTED READING

Angier, Bradford. *Field Guide to Edible Wild Plants.* Harrisburg, PA: Stackpole Books, 1974. This is an excellent slim book for identifying wild foods. Fully illustrated with color drawings, most of which are fairly accurate. Covers the entire United States.

Coon, Nelson. *The Dictionary of Useful Plants.* Emmaus, PA: Rodale Press, 1974. A well organized book dividing over 600 plants into families, including several lists of references. Weak point is lack of illustrations.

Gabriel, Ingrid. *Herb Identifier & Handbook.* New York: Sterling Publishing Company, Inc., 1977. A lovely little color-illustrated hard bound book; includes where herbs are found throughout the country, plus medicinal uses.

Gales, Donald Moore. *Handbook of Wildflowers, Weeds, Wildlife and Weather of the Palos Verdes Peninsula.* San Pedro, CA: Caligraphics Printing and Publishing, 1974. Can be used over most of the western United States, and to a limited extent, throughout the United States.

Grieve, M. *A Modern Herbal.* New York: Dover Publications, 1971. 2 vol. The medicinal and cosmetic properties, cultivation and folklore of herbs, grasses, fungi, shrubs and trees with modern scientific uses.

Harrington, H.D.; Durrel, L.W. *How to Identify Plants.* Chicago: Swallow Press, 1957. The best single book for plant taxonomy. Learn all the parts of a plant. The heavily-illustrated book is divided by plant parts and alphabetically.

Harrington, H.D. *Edible Native Plants of the Rocky Mountains.* Albuquerque, NM: University of New Mexico Press, 1974. A thick manual combining scientific accuracy with straight talk.

Jacobs, Betty E. M. *Growing Herbs and Plants for Dyeing.* Tarzana, CA: Select Books, 1977.

Kirk, Donald R. *Wild Edible Plants of the Western United States.* Healdsburg, CA: Naturegraph Publishers, 1970. Over 300 pages describing identification and use of 302 plants, with reference to many more. The book is strong in its attention to so many plants, divided by location, but it is weak because it tries to cover too many. Because only basic identifying features are given, it may be dangerous to use this book as an only reference.

Knobel, Edward. *Field Guide to the Grasses, Sedges and Rushes of the United States.* New York: Dover Publications, 1977.

McD. Schetky, Ethel Jane; Woodward, Carol, eds. *Dye Plants and Dyeing—A Handbook.* Baltimore: Brooklyn Botanic Garden, 1976 (July) Vol. 20, No. 3. Published quarterly.

Menzies, Robert. *The Herbal Dinner.* Millbrae, CA: Celestial Arts, 1977. A beautiful book that helps identify wild food and offers recipes for using this free food supply.

Nyerges, Christopher. *A Southern Californian's Guide to Wild Foods.* Los Angeles: White Tower Inc. Press, 1978.

Silverman, Maida. *A City Herbal.* New York: Alfred A. Knopf, 1977. A guide to the lore, legend and usefulness of 34 plants that grow wild in the city, with recipes for breads, salads, seasonings, teas, dyes, cosmetics, potpourris and more.

Storer, Tracy; Usinger, Robert. *Sierra Nevada Natural History.* Los Angeles and Berkeley: University of California Press, 1963. This book was used as a textbook in many college biology departments. It is an excellent introduction to geology, history, flora and fauna of the Sierra Nevadas.

Wilkins, Marilyn. *California Dye Plants.* Santa Rosa, CA: Thresh Publications, 1976.

WATER PLANTS

I became interested in wild foods during my many hiking trips throughout the Angeles National Forest in Southern California. At a place called Inspiration Point, I met a curly-haired, bearded man who was doing handstands on the edge of a steep precipice. We talked for a few moments and I learned that he was trained in Indian lore, specifically the foods the Indians gathered and ate. He mentioned some of the wild foods he had eaten that day. Although we talked for only a few moments, that conversation was the spark that took me all over the United States studying wild foods.

I lived in the Eastern part of the country for a while and became familiar with the edibles in that part of the country. Trips throughout the Southwest and throughout the Northwest increased my awareness of our nation's rich bounty of free food. Eventually, I began teaching wild food classes and leading wild food hikes to teach others how to identify and use these wild plants. Some of my most enjoyable outings have been along the coast, learning, searching, gathering and sampling the vast array of delicious foods that are available to any peripatetic vagabond who travels these salty shores.

A broad array of edible plants flourish in coastal areas, on sandy beach dunes, in salt marshes and around waterways, in general.

ASPARAGUS *(Asparagus officinalis)*

Although not unique to the coastal areas, wild asparagus grows abundantly in salt marshes and bay areas. When mature, the plant looks like an ornamental fern with small red fruits that develop in the fall. Gather this plant in the early spring, when it sends up new shoots; it is identical to the asparagus you buy in the produce store. Asparagus is delicious boiled, steamed, creamed in soups and added to vegetable dishes. You can enjoy asparagus shoots raw, but some people develop a type of dermatitis after eating the raw shoots. I suggest you go slow with raw asparagus, until you are certain your body can handle it.

CATTAIL *(Typha latifolia)*

Cattail grows throughout the United States in swamps and marshlands everywhere. Its long grasslike leaves and annual brown, sausage-like flower stalk makes cattail easy to recognize. Cattail is a perennial plant. The leaves grow up to five feet in length and are an inch or two wide. The root stalk, lower leaf stem, young shoots and fruiting flowerheads are all highly usable for food.

Before the flower stalk rises in the spring, the tender core of the leaves is delicious raw in salad. In the spring, peel back the outer fibrous dark green leaves from the base of the stalk and pull out the white tender core of the stalk. The taste is similar to raw cucumber. As the plant sends up a flower stalk, the inner core becomes tough and too fibrous to eat. Tug briskly on the central core as a slow pull will usually bring up the root of the plant. This means you've got an unnecessary cleaning job on your hands.

Don't fret if you accidentally pull up the roots. The starchy section at the base of each stalk, if the plant is still young, can be eaten raw or cooked. The texture is similar to a potato. The Iroquois dried the rhizome roots for a sweet flour from which they made bread and pudding. Cattail flour is highly nutritious.

Gather the cattail flower spike in the spring, while it is still green. Most of the flower stalks will be fully mature by summer. Either steam or boil the green flower spikes. Butter, season and eat like corn-on-the-cob. You will be amazed at how similar cattail and corn taste.

CATTAIL CAKES

¾ cup cattail pollen
1½ cups whole wheat pastry flour
1 egg
1 cup milk
1½ Tbsp. honey

Mix ingredients together in order listed. Thin or thicken to desired consistency. (A thin batter will make excellent pancakes.) Bake in a 350° oven for ½ hour.

Cattail provides many essential survival items: food, fibre, insulation, fire-starting tinder, and potential shelter. The long slender leaves of cattail make an excellent weaving material for sandals, mats, lightweight baskets and many other items. Although the leaves can be used green in an emergency, they are stronger and last longer if they are picked green, allowed to dry in the sun for two to five days, and then soaked in water to be made pliable before using. There are many methods for weaving fibres together. Two of the most simple are braiding and the "over and under" technique.

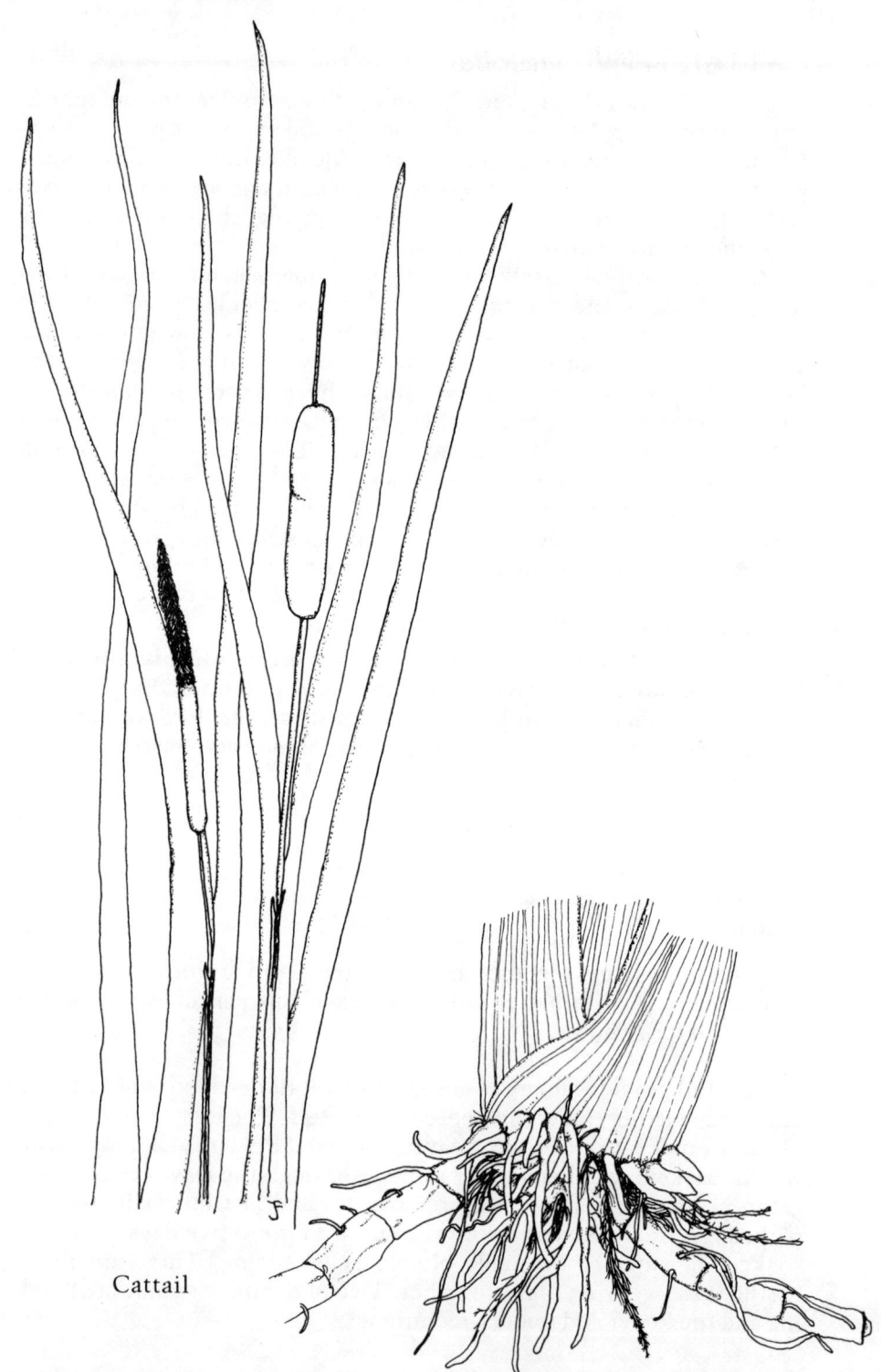

Cattail

When the fully mature brown flower spikes are broken open, the compacted fibres will freely disperse, revealing an abundance of good-quality stuffing material, insulation, or fire-making tinder. If you decide to use this material on cold nights, be sure to have something into which to stuff all the fluff. Otherwise, you will be breathing and sneezing cattail all night.

GLASSWORT *(Salicornia sp.)*

Glasswort is found in salt marshes slightly above the high-tide line from Mexico to Canada on the West coast and from Florida to Massachusetts on the East coast. It is a succulent plant whose stems appear as opposite, plump, segmented pencillike divisions. These succulent stems have small scaly leaves at each node, but you will have to look closely to see them; the entire plant seldom reaches two to three feet high. When the plant turns a beautiful shade of red in the fall, it is no longer worth gathering for food. While still young, it adds a deliciously crisp, salty taste to your salad. Pinch off the upper and more tender sections of the plant.

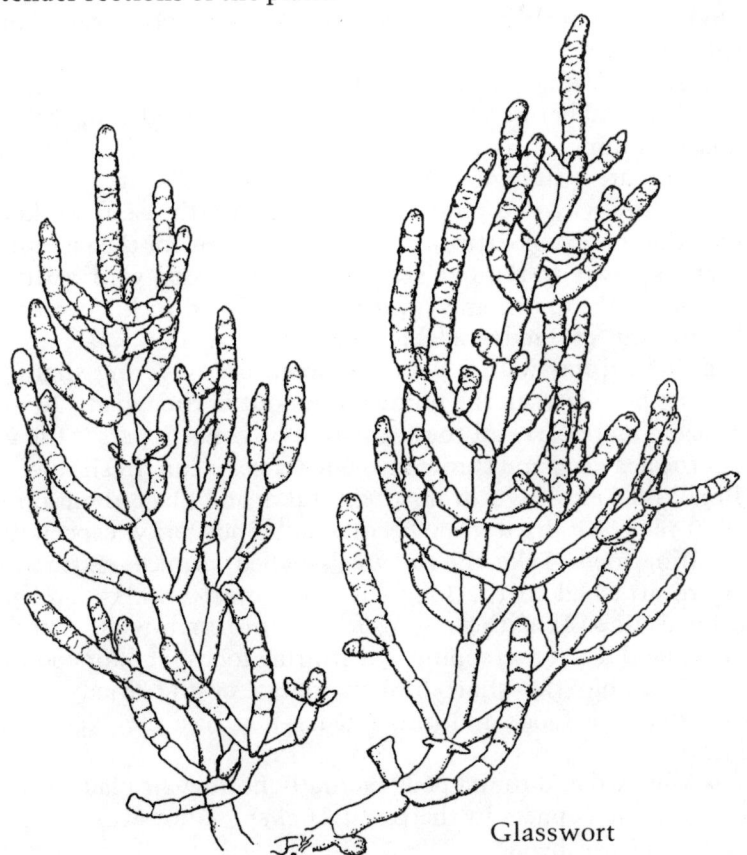

Glasswort

KELP

Anyone who's visited ocean beaches has encountered kelp and mistaken it for a nuisance. What could be more repulsive than those gooey, smelly clumps of fly-infested seaweed? The most outstanding impressions are the strong "oceany" smell and its sliminess. With food gathering so removed from most people's lives, it is no wonder we overlook seaweed and its many potential uses in our daily lives. Once you lose your prejudice for the slime, you develop a respect for the usefulness of the various seaweed parts. Seaweeds should not be overlooked, as they are a rich source of many vitamins and minerals and are so readily available.

Sea lettuce *(Ulva lactuca)*, a bright-green translucent seaweed is found in thin sheets and ribbons like tissue paper. Usually attached to rocks, it is observed most easily at low tide. Chop sea lettuce fine and add it to salad and other cooked foods. Dried and powdered, sea lettuce becomes an excellent seasoning and salt substitute.

Two other superior seaweeds are dulse *(Rhodymenia palmata)*, wine to dark red colored with tongue-shaped divisions, and laver *(Porphyra sp.)*, purplish-brown with either fronds or a series of lobes and ruffled edges.

All seaweeds are edible. Eating them raw is sometimes like chewing on a rubber band, but once dried and powdered, they can be used as seasoning or as a soup base. Dried they taste like seaweed jerky; many are quite good. Some, as you will discover, really don't taste good any way you prepare them. Be sure to use fresh kelp (not seaweed that has been sitting on the beach rotting for the last month).

Here is a way to remove seaweed slime. Using your bathtub or a large sink, wash, drain, and rewash the seaweed in hot water with a small amount of biodegradable soap. Repeat the wash and drain process three times with a water temperature hovering slightly over 110° (most hot water in the home comes out of the tap at 120°-140°). All the sand and nearly all the slime will be gone. Separate the various parts of the seaweed and throw into buckets of clean, fresh rinse water.

Break off kelp leaves from their stalks and place them on a wire screen or rack. Air dried, they become delicious jerky, especially good as travelling food. Take the flat oval-shaped, crinkle-textured leaves that were attached to the long succulent round stalks, and prepare them for drying. Tear these long leaves into inch-wide strips. Later remoisten and use for wrapping and stuffing foods, and for cooking and baking. Crumble the short broken pieces of the dried leaves and sprinkle over hot foods and into salads—just like the expensive kelp from health food stores.

Use the oval and round pods (actually hollow air bladders) as substitute for chile peppers in the traditional spicy-hot Mexican preparation, chiles en escabeche.

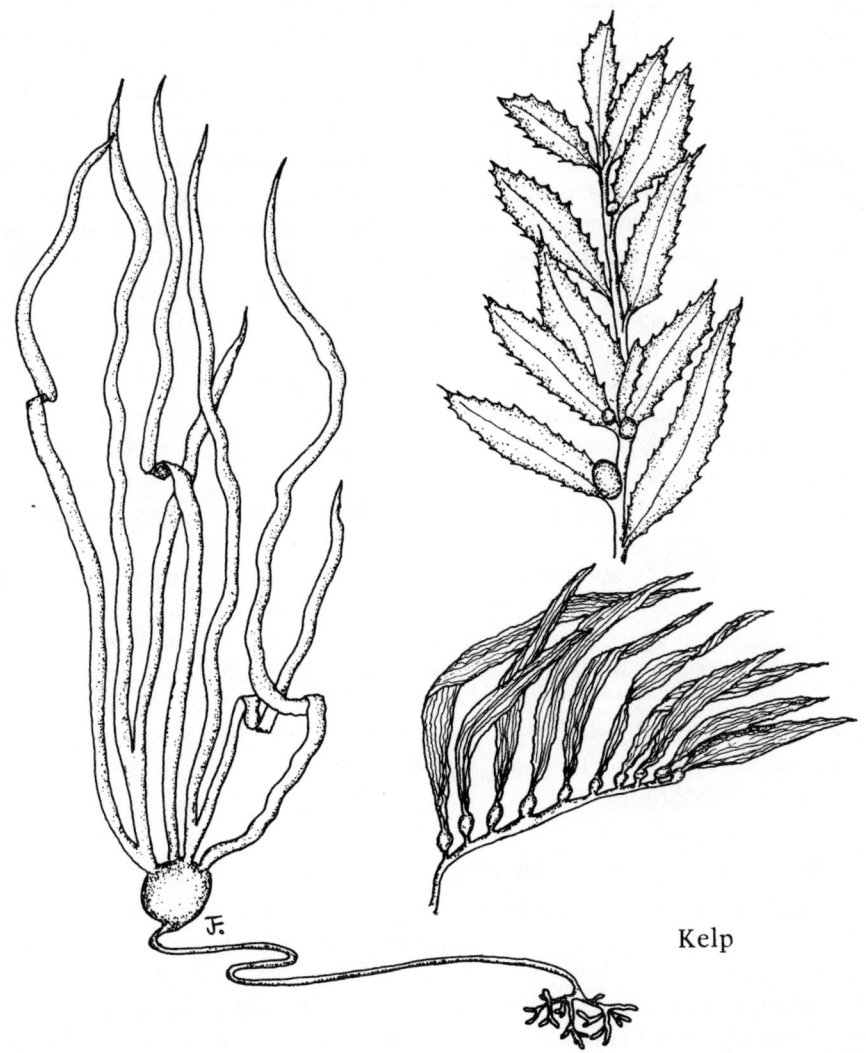

Kelp

PICKLED KELP

2 one quart glass jars
Approximately 100 kelp air
   bladders
Apple cider vinegar, to cover
4 Tbsp. olive oil

Sprinkle tarragon
Sprinkle dill seed
6 cloves fresh garlic, sliced
Dash celery powder

    Put the raw pods (alone or with other vegetables of your choice) into clean quart jars. Add apple cider vinegar until the pods are nearly covered, and top with a quality cold-pressed olive oil. Sprinkle with dill seed, tarragon, sliced fresh garlic cloves and celery powder. Cover tightly and shake occasionally for a few days. The mixture will be ready to eat as is, and can also add a bit of nutritious "Wow" to sandwiches, salads, omelets and Mexican dishes.

Both the flat and the round stems and stalks from seaweed are the basis of yet another group of products. Break into 1" and 2" lengths, put into a large covered kettle filled with spring water (or, if you're able to get it, clean seawater). In a blender mix into a mush. Pour mixture through a fine-mesh metal strainer or one layer of muslin cloth to strain out solids. Bottle, label and refrigerate the resulting liquid and use as a gravy or soup base. Use for a slightly oceany-smelling "mineral bath," or as a laundry water softener which will enable you to cut back 30%–40% on soap.

The strained-out pulp has many uses: cooked into homemade ice cream blends, it is a peerless smoother/stabilizer; as worm food (if you have a worm bed); as compost/mulch for the garden; dried and added to animal feed; as a substitute for flours in making smooth gravies, sauces and soups.

Nonfood uses should not be left unmentioned. Although these applications might seem reserved for the island adventurer or the shipwrecked sailor, sheer ingenuity impels me to share them. The long flat stems can serve as an interim material for lashing/binding (preferably in places where they will not get wet). These same stems, along with the many seaweed "strings" you will find mixed in with the seaweed clumps, can be used for making moccasins, clothing repairs, mats, baskets and pot holders.

LETTUCE, MINER'S *(Montia perfoliata)*

Miner's lettuce is an excellent wild food found throughout the Western states from the ocean to the mountains. Miner's lettuce grows prolifically along the Pacific Coast, throughout the Sierra Nevada mountain range and scattered throughout the Southwest.

Miner's lettuce is easily recognized by its circular cup-shaped leaves which encircle each flower stalk. The entire plant grows about eight to ten inches tall with white to pink, small five-petalled flowers. The basal leaves of this plant are either triangular or rectangular in shape.

Each plant can be picked easily, and the basal section somewhat resembles the base of loose-leaf lettuce, which gave rise to the plant's name. The plant earned the adjective "miner's" because the California '49ers ate large quantities of this plant to prevent scurvy. Miner's lettuce is an excellent source of both vitamin C and iron.

Miner's lettuce is best eaten raw or steamed slightly. It is delicious in salads, mixed with hard-boiled eggs, tomatoes, onions and

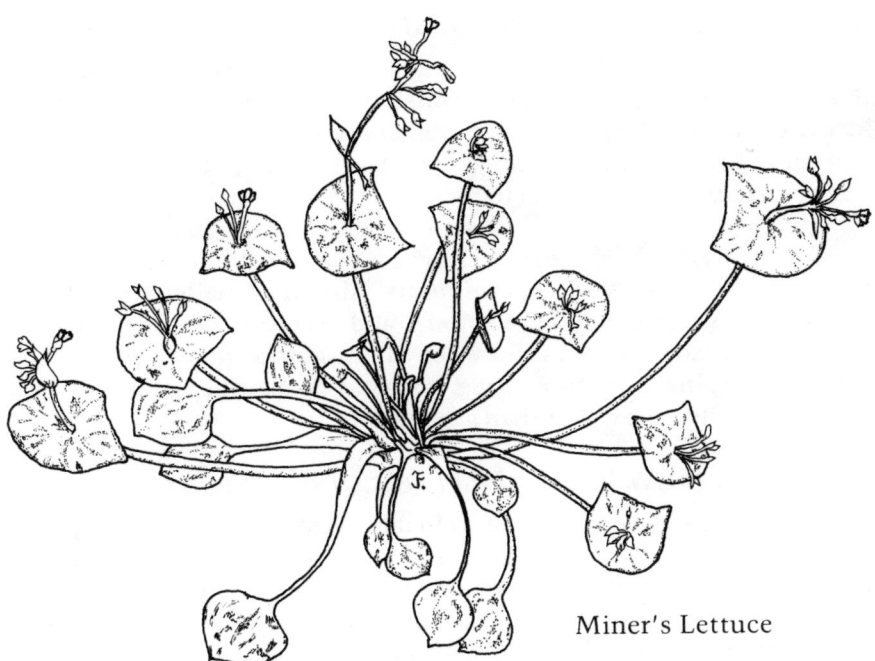
Miner's Lettuce

avocado. Eat the entire plant. Season simply with a vinegar and oil dressing for a fine-tasting salad. Steam miner's lettuce, and season with butter.

HERB SALAD

1 cup miner's lettuce
½ cup dandelion greens
½ cup plantain leaves
1 cup Romaine lettuce
½ cup alfalfa sprouts
½ cup chopped parsley

Wash and dry the greens. Tear into bite-size pieces. Mix with sprouts and chopped parsley. Excellent with green goddess or watercress dressing.

ORACH *(Atriplex sp.)*

Orach is found from British Columbia to Southern California on the west coast and from Labrador to Virginia on the east coast. Orach with its triangular-shaped, tooth-margined, mealy covered leaves, is a close relative of the common mainland weed, lamb's quarters *(chenopodium album)*. Its hastate leaf shape (the basal lobes of the leaf sharply point outward at a 90° angle from the leaf's central axis) is most characteristic. This common beach plant is "presalted" and rich in iron, vitamins, and seasalt minerals. Cook it as you would spinach or eat it raw in salad.

SEA ROCKET *(Cakile edentula)*

Sea rocket, an annual member of the mustard family, is found in the sand on both coasts. It has purplish or white four-petalled flowers. Identify sea rocket by its fleshy, oblanceolate to obovate leaves with sinuate margins, and plump and green two-valved seedpods. Sea rocket has a pungent horseradish taste.

In late autumn you will find a large and bushy plant already gone to seed and dying. Under the plant where the seeds drop, you may find hundreds of tender sea rocket sprouts, slightly smaller than alfalfa sprouts. I have gathered these sprouts and used them in many dishes, especially salad and clam chowder. The mature plant, even when boiled like spinach, is too strong for most people. Sea rocket is best used as a condiment rather than a staple.

SPINACH, NEW ZEALAND *(Tetragonia expansa)*

New Zealand spinach is a valuable west coast shoreline plant. Well known to many, it is commonly found in the markets of certain locales. It has a similar appearance to spinach, although its rhombic-ovate leaves are covered with a sparkly crystalline substance. Cook as you would spinach and season with butter, or add it to other cooked sea foods. It is delicious eaten raw in salad. New Zealand Spinach can be found along most of the California Pacific Coast.

WATERCRESS *(Nasturtium officinale)*

One of the best places to look for watercress is where inland streams empty into the ocean. Look also in slow-moving water, such as ponds, lakes and edges of streams, especially in the spring. *Avoid picking watercress where the water has any possibility of being polluted.* This member of the mustard family, found throughout the United States, is quite versatile. Perhaps you have seen it in produce sections. Eaten raw, the young pinnately-divided leaves make a salad spicy and rich in Vitamins C, E and other minerals. Eat the entire above ground plant.

Xenophon, an ancient Greek historian and general, and Xerxes, a Persian king, observed that those who ate watercress were invariably in better health. Xenophon recommended his soldiers include watercress in their diet.

SAUTEÉD WATERCRESS

1 cup chopped onions
2 Tbsp. butter
10 cups watercress
Tamari or soy sauce to taste

Fry onions lightly in butter. When the onions are slightly browned, add watercress. Do not overcook; simply let the leaves wilt before serving. Season with tamari or soy sauce.

WATERCRESS SOUP

2 cups milk
1 cup chopped watercress
Paprika to taste

Heat the milk (do not boil). Add watercress and cook for five minutes. Season with paprika and serve with a sprig of watercress. Expect many compliments on this toothsome soup.

SUGGESTED READING

Christopher, F.J. *Basketry.* New York: Dover Publications, 1952. This is a slim, but excellent book on all aspects of basketry, including materials needed and various types of construction. Makes reference to cattail for basketmaking.

Kirk, Donald R. *Wild Edible Plants of the Western United States.* Healdsburg, CA: Naturegraph, Inc., 1975.

Madlener, Judith Cooper. *The Seavegetable Book.* New York: Clarkson N. Potter, Inc., 1977. Details the foraging and cooking of seaweeds.

Rhoads, Sharon Ann. *Cooking with Sea Vegetables.* Brookline, MA: Autumn Press, 1978.

CACTI

We have all heard that some cacti are edible, and that lost desert wanderers have been saved by obtaining food and water from certain species of cacti. City residents in the Southwest, Arizona, New Mexico, Texas and some parts of California find common cacti growing wild.

Flat oval-shaped fleshy pads, bearing numerous spines and fuzzy hairlike glochidia, make prickly pear easy to recognize. Comes spring and prickly pear flowers in beautiful shades of yellow, orange and red. Prickly pear provides a pleasant break in an otherwise drab desert area. As summer draws to a close, the pears develop fully and ripen to red, purple or yellow.

Prickly pear is one of the most abundant and easiest cacti to use for food and drink. The ripe fruit are delicious and juicy. The inside is usually red (sometimes yellow), very seedy and has a taste and texture similar to watermelon. Prickly pear has quenched my thirst many times. *Exercise caution when gathering this fruit;* it is covered with minute glochidia (barbed tips) which are very irritating once they enter the skin. Try pulling on the fruit with a bag around it. You will have the fruit in the bag, and avoid contact with the glochidia.

When you're ready to eat it, grab the butt end of the fruit with a fork, scrape off all the glochidia with a small knife and rinse. Cut both ends of the pear, but not all the way off. Cut a third line perpendicular to the first two cuts, so that you've cut an *H* into the fruit. The fruit will resemble the individual cereal box so popular with children. Peel back the skin and pop the prickly pear into your mouth. Ways to use the fruit include drinks, candy, jams, ice cream and jelly.

PRICKLY PEAR MILK SHAKE

3 cactus pears, cleaned and diced
1 apple, diced
1 pear
1 banana
1 cup chopped nuts
2 cups milk
2 Tbsp. honey

Put ingredients into a blender and mix until smooth. Chill before serving.

The fruit is the most well known edible part of this plant, but the young pads or nopalitos are also edible, and are available all year long. Pick nopales when they are small (nopalitos) and glossy-green. Exercise caution when gathering the pads also. Pick the young pads carefully, shielding the hands with a bag, or scrape the spines free of glochidia with your knife before you pick the pads. Each pad must be cleaned of all spines and glochidia before it is eaten. Once the skin is peeled, dice and pad into bite-size pieces. Nopalitos are good in virtually any salad mixture.

CACTUS SALAD

1 cup diced cactus pads
2 cups diced tomatoes
1 cup diced celery

1 small onion, finely chopped
1 green pepper, chopped

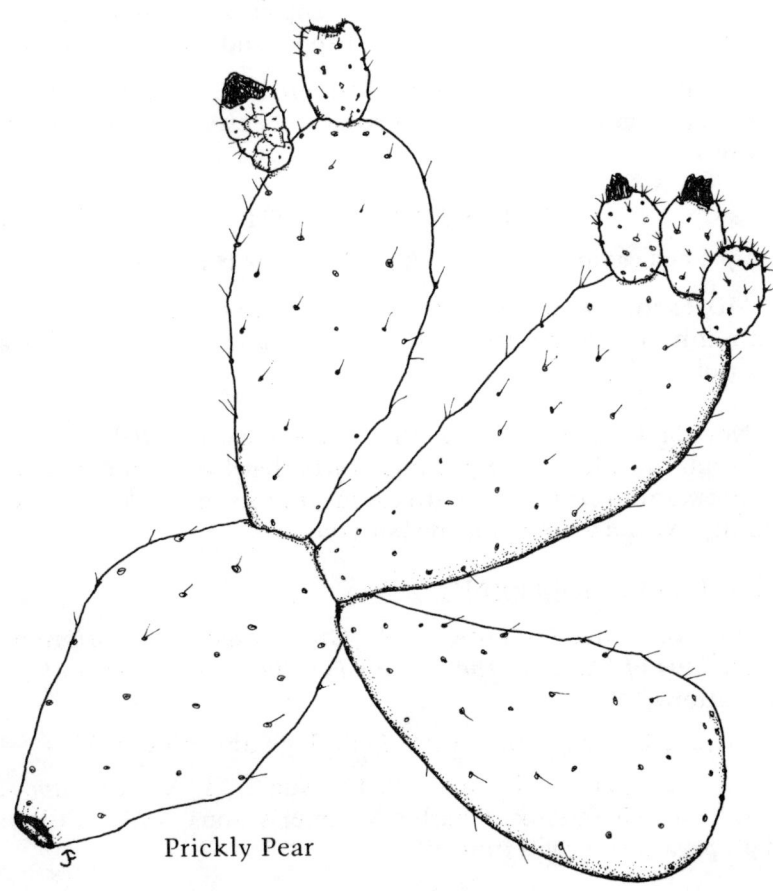

Prickly Pear

Mix all ingredients. Season with your favorite salad dressing, or use a dressing of olive oil, wine vinegar and a teaspoon seasoning salt.

NOPALITO OMELETTE

2 cups peeled and diced cactus pads
1 onion, diced
4 eggs, beaten
1 Tbsp. butter

Sauté diced cactus pads and onion in butter until the juices of the cactus are released. Continue cooking until the juices evaporate and the cactus has turned from bright green to light brown. Add eggs and scramble. Serve on a hot tortilla.

NOPALES POTATO CASSEROLE

4 potatoes, boiled and mashed
5 young pads, peeled and finely diced
½ cup butter
2 eggs, beaten well
1 onion, minced
1 cup grated cheddar cheese
Pepper and dill weed, to taste

Mix potatoes and diced pads. Add butter, eggs and onion. Season with pepper and dill weed. Cover with cheese. Bake at 250° for 40 minutes.

CACTUS HAIR CONDITIONER

½ cup diced cactus pads
2 cups warm water

Mix cactus, water and stir. A very gelatinous mixture will result within minutes. Strain the cactus chunks, and use this juice as a hair conditioner.

Nevada Indians believed that prickly pear cured warts. They would rub the juice of the pad into warts that had been cut open. You may not want to cure warts with cactus, but its multiple food uses will certainly expand your mealtime horizons.

SUGGESTED READING

Britton, Nathanial L.; Rose, J. V. *The Cactacae: Descriptions and Illustrations of Plants in the Cachos Family.* 2 vol. New York: Dover Publications, 1937.

Herbert, Frank. *Dune.* New York: Berkeley Publishing, 1975.

Martin, Margaret J.; Chapman, P. R.; Auger, H. A. *Cacti and Their Cultivation.* New York: Charles Scribner's Sons, 1971. Cultivation and propagation of cacti from all over.

Organic Gardening. "Ancient Crops for Desert Gardens," February, 1977, p. 34.

Sunset, "Good Looking . . . Unthirsty," October, 1976, p. 80.

Venning, Frank D. *Cacti.* New York: Western Publishing Company, Inc., 1974.

FOOD GROWS ON TREES

Trees provide an excellent source of free wild food. Walking down some city streets, you can gather foods that are generally expensive in the stores: nuts, avocadoes, carob, olives and fruit. Study your local trees and investigate the many uses of these nutritious and delicious foods. Gradually you will develop the forager's eye.

NUTS

For generations nut collecting has been an autumn ritual in many parts of the United States. Besides saving money, nut gathering outings are an excellent time for the family to get together to talk, laugh and work. Some of the most common native North American nut trees are the black walnut, butternut, hickory, chestnut, beech and pinyon.

The black walnut *(Juglans nigra)* is found throughout the Eastern United States and in parts of Southern California. The black walnut of the Eastern United States is a magestic tree, often reaching up to 100 feet. In California the black walnut *(J. californica and J. hindsii)* does not get as large, nor as beautiful. It grows to a height of 30 to 60 feet.

The green nuts begin to drop in October and November. Take care when gathering, because the husks contain a fairly long lasting dark brown dye. You may want to wear gloves for protection. Crack the hard shells with a hammer and separate out all the meat. If you think this sounds like a rather tedious job, you are right. Why bother? Because the nut meat is delicious and can be used just like you would use regular walnuts. Use black walnuts in any recipe calling for walnuts. They are great cooked into wheat bread, cookies, granola and used as a topping (once chopped) for cakes, pies and ice cream.

The butternut *(Juglans cinerea)* is another wild walnut tree, smaller than the black walnut and with a slightly lighter bark. Butternut is also found further north and in higher elevations. The butternut is elliptically shaped, rather than the sphere shape of the walnut. Their meat is even harder to extract than the black walnut, but some cooks prefer this delicious oily meat. Immature butternuts are also good pickled. The process varies from location to location, but here is one recipe. To test butternuts, to be certain they are not too old to pickle, use a nutpick or icepick to pierce the husk and shell. Collect the pale-green half-grown butternuts, pour boiling water over them, and then rub off the fuzz from the outside of the husks. Soak them in a strong brine and change this liquid every day for a week. Drain the nuts, and pack them into quart jars, adding dill, pickling spices, and salt. Fill the jars with boiling cider vinegar and seal. They should be ready to eat in a about a month.

The hickory tree *(Carya sp.)* is found primarily in the Eastern United States. They are large, well-crowned trees, rising up to 70 feet high. Hickory and walnut trees are related. Walnut leaves are compoundly-divided with from 15 to 21 leaflets, all in pairs except the terminal leaflet. Each leaflet is from four to six inches long, lance-shaped and tapered at the point with serrated edges. Hickory leaves are compound also with five to seven leaflets. The inner two (or sometimes four) leaflets are smaller than the outer three. The hickory nut differs from the walnut in that the husk of the hickory divides into four segments, allowing the nut to fall free easily.

Some hickories are not used for food simply because they don't taste good. Most, however, are edible. Break one open with a hammer; if you find it agreeable, gather a bushel and take them home. People who know their nuts say the shagbark hickory *(Carya ovata)* has the best tasting nuts of all the hickories.

The pecan tree is a type of hickory that prefers the climate of the South and Southwest but has been planted in the northern states, because it is an impressive ornamental. Wild pecan shells are thicker than commercial varieties, but are well worth the extra effort.

The pinyon pine *(Pinus sp.)* is a small shrubby tree found throughout the Southwest. Although the nuts of all pines are edible, pinyons are best. These delicious nuts were once highly prized by the Southwest Indians. Nuts were generally collected after the pine cones opened and released the seeds, but sometimes the unopened cones were gathered and opened. Each seed is encased in a hard black thin shell which can be opened with the teeth or hit with a rock or hammer. Pine nuts or seeds are good raw, but they can also be roasted, baked into breads and added to desserts. They are an excellent food for the pineal gland, and are 15% protein, 62% oil and 17% other carbohydrates.

ACORNS *(Quercus sp.)*

Begin your wild food study with plants you already know. The oak tree is certainly a good start, because they are found throughout the

United States. In September and October they are identified easily by their fruit, the acorn, which falls to the ground in great numbers.

Acorns were the staff of life to early Southern California Indians. These Indians had several different ways of preparing acorns, and they stored up to 500 pounds (both shelled and unshelled) for a family's yearly supply. The acorn is 65% carbohydrates, 18% fat and 6% protein.

The presence of tannin in the raw acorn makes it very bitter. A number of methods have been devised to rid acorns of their bitterness. Indians shelled acorns and put them in a cloth overnight in the river. By morning the flowing water would leach the water-soluble tannin from the acorns. Another Indian method: Place ground acorns into a cloth or burlap-covered bowl, made out of twigs or pine needles, supported about 2 feet off the ground by vertical stakes. Pour water into the acorn meal, and allow it to filter through. Leaching time depends on the bitterness of the acorns, but a few hours is usually sufficient. The final product can be boiled up into a mush, and eaten hot or cold.

Boiling is the quickest way to make acorns edible. Shell the acorns and boil, changing the water as it becomes brown. They are ready to eat when the bitterness is gone. Once sun dried or slowly oven dried, grind them with a hand or electric grinder, or stone metate. Use the flour in bread, muffins, pancakes, grits and soup, mixed with wheat or corn flour. Indians mixed the acorn meal with water and ashes and baked it into a flat bread in crude ovens. Indians often mixed cornmeal with acorn meal together in cooked foods.

ACORN BREAD

1 cup acorn meal
¾ cup whole wheat flour
¼ cup carob flour
3 tsp. baking powder
1 tsp. sea salt

3 Tbsp. honey
1 egg
1 cup milk
3 Tbsp. oil

Mix ingredients well and cook in a greased pan for about 30–45 minutes at 300°.

Use of the oak doesn't stop at the acorn. The bark has medicinal properties and is used in the tanning of buckskin. Acorns provide an orange-tan dye which takes especially well to wool. Boil acorns and simmer (an hour a pound); then strain and use. Excellent dyes of various colors can be obtained from mixing oak bark with other plants.

CAROB *(Ceratonia siliqua)*

Today it is nearly impossible to avoid white sugar in any processed food. Not more than a few centuries ago one of the major food sweeteners in the world was a type of healthful "chocolate" that grows on trees. It is believed that the fruit of the carob tree helped feed Mohammed's armies, John the Baptist during his sojourn and meditations in the wilderness, and the prodigal son who was hungry and without money. Children during the Spanish Civil War who ate carob remained free of malnutrition. As recently as World War II isolated military troops and their horses on the island of Malta and people in Greece credit their survival during the German occupation to the carob fruit. Carob is just now, slowly but surely, regaining its once important role in the diet.

The fruit of the carob tree is a dark brown, flattened leathery pod (or legume). Carob is native to countries surrounding the Mediterranean Sea. The best commercial carob comes from this area; propagated from root stocks, the trees produce superior carob fruit. Southern Californians and Arizonians know the ornamental carob trees that are so widely planted as street and park trees. These showy evergreens are drought-resistant, their long roots reaching deep for underground water. Each carob leaf is pinnately divided into 6 to 10 round, glossy, leathery leaflets. Each pod measures from 1 to 2 inches wide and 4 to 10 inches long.

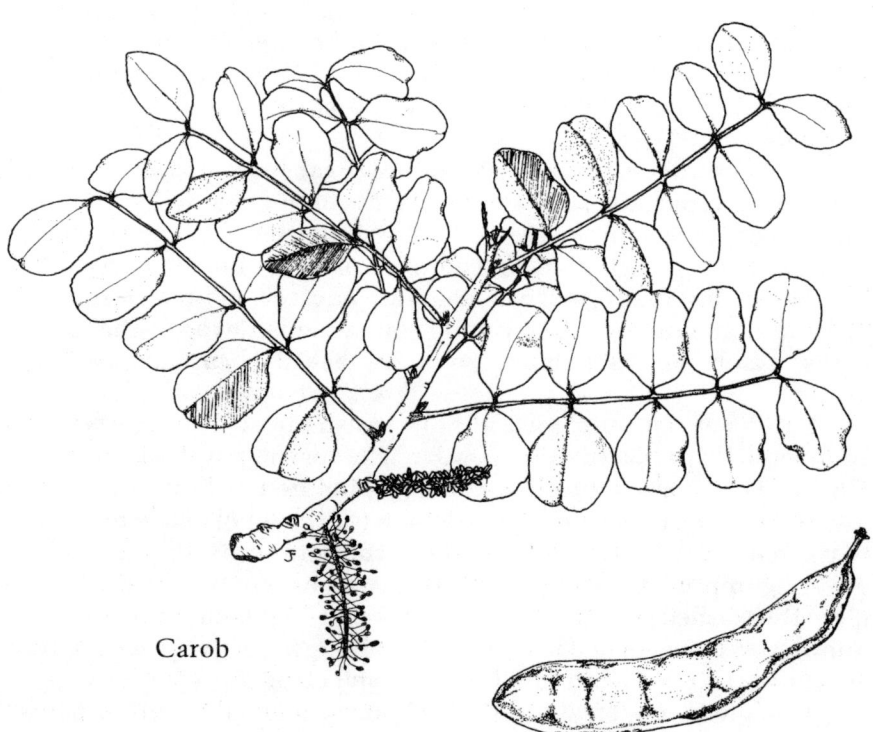

Carob

Carob contains about 4% protein and 76% carbohydrates. Although it is very sweet, carob has 60% fewer calories than chocolate. Carob contains substantial phosphorus (81 mg. per 100 grams or 367 mg. per pound), an abundance of potassium (800 mg. per 100 grams), and small amounts of sodium and iron. Carob is particularly rich in calcium (352 mg. per 100 grams or 1,597 mg. per pound). Milk, which we consider an excellent source of calcium, contains only 120 to 130 mg. calcium per 100 grams, or 530 to 550 mg. per pound. Because carob has no oxalic acid—chocolate does—it will not interfere with the body's assimilation of calcium. Carob lacks caffeine and contains only traces of theobromine, the active stimulant of chocolate and cocoa.

Eat carob pods right off the tree. The sweet and chewy pods make nutritious snacks. Once dried, whole carob can be stored for future use and carried along on trips. Carob powder is available from many sources. Try purchasing it in bulk from your local health food store or market. Be sure you are getting pure, unadulterated carob powder; check the label. Sugar, chocolate and/or cocoa is occasionally added to commercial carob powder.

Use carob powder whenever a recipe calls for chocolate or cocoa. Carob is so drastically different nutritionally and chemically from chocolate that people allergic to chocolate can enjoy carob. Use the same amount of carob as you would cocoa. To replace chocolate, use approximately three tablespoons of carob powder for each square of chocolate in a recipe.

Carob powder is almost 50% natural sugar and can be used instead of sugar in bread and pastry products, including waffles, cakes, pie, pancakes, cereals (hot or cold), crepes and muffins. Carob powder will turn all of these foods chocolatey-brown and will impart a chocolate-like flavor. Try mixing various amounts of carob and honey to find a mixture that suits you. Use carob for its unique flavor in shakes and malts, carob nut bars, bread products, baked beans and barbeque sauces.

Carob powder purchased in bulk is usually slightly coarser than commercially packaged carob powder. The coarser powder is similar to what you can make yourself, gathering, grinding and drying the pods.

Western readers who have a source for the whole pods may want to try making their own flour. Remove the hard seeds; they are notorious for gumming up grinders. An easy way to remove the seeds is to place the washed pods in a pressure cooker at 15 pounds pressure for 20 minutes. When cool and dry, the pods split open easily for seed removal. Cut the soft pods into small sections and blend until powdery.

You might try breaking the pods open manually with a pair of pliers. Once the seeds are removed, dry the pods slowly in the oven at a very low temperature. Grind the seeded pods in a stone grinder at a coarse setting. Dry the coarsely ground pods again for a few hours and regrind at a finer setting.

I have used coarse carob powder as a flour substitute. As the only flour ingredient, carob flour will make a flat bread. For best results mix half carob flour and half wheat, corn or rice flour.

HONEY CAROB BROWNIES

1 cup wheat flour
1 tsp. baking powder
¼ tsp. salt
½ cup butter
½ cup carob powder

1 cup honey
2 eggs
½ cup chopped walnuts
1 tsp. vanilla

Sift together flour, baking powder and salt. Melt the butter in a small pan over low heat. Add carob powder, honey and blend well. Remove mixture from heat. Beat eggs in a mixing bowl and gradually add the carob mixture. Add dry ingredients and mix well. Blend in vanilla and nuts. Pour into an oiled 8 inch pan, and bake at 300° for about 40 minutes (or until done).

CAROB SAUCE

5-8 Tbsp. carob powder
3 Tbsp. honey

Cream or milk

Pour a little milk or cream and honey into a blender. When mixed well, slowly add carob powder. Add more milk or cream as the mixture thickens. When the carob is thoroughly dissolved, add milk or cream to reach the desired consistency. Use over ice cream, in malts and shakes, or in candy.

CAROB SHAKES

3 scoops vanilla ice cream
1 egg
2 cups milk

1 Tbsp. carob powder
3 Tbsp. carob sauce
1 banana

Blend ingredients well and enjoy.

Carob is also widely known and used as a medicine for the treatment and prevention of diarrhea in livestock. Only recently have medical journals in North America and Europe begun advocating carob powder for the prevention and cure of human dysentery, especially in children. The pectin and lignin in carob not only regulate digestion, they combine with harmful elements—even radioactive fallout—in digested food and carry them safely out of the body.

Marian Seddon
Desert, December, 1972

Carob pods have been widely advocated in the United States for use as horse, goat, pig, cow and poultry feed. My goats crave carob pods and eat all I can supply. Carob is an extremely beneficial food that we would all do well to include in our diet.

OLIVES

Olives are a delicious and nutritious food, but unfortunately we cannot eat them right off the tree due to their astringency. All too often the fruit is not used, because many people think processing olives is too complex. How I fret when I see these wonderful fruits falling to the ground, stepped on and swept away.

Olive trees are a shade and park tree in many parts of the United States. If olive trees grow in your vicinity, become familiar with a few basic methods for processing olives.

GREEN OLIVES

This process will produce straw-yellow to green or brown olives. Choose ripe green, straw-colored or cherry-red olives. I do not recommend using black olives with this method, because when pickled, they are likely to become soft.

Dissolve a twelve-ounce can of flake lye (household lye) in six gallons of water. (Two ounces or four tablespoons lye to each gallon of water.) Use a wooden, glass or stoneware container; never use aluminum (lye will ruin it) or galvanized metal (the zinc will dissolve and may make the olives poisonous). Stir the solution with a wooden or stainless steel spoon until the lye is well dissolved. Let the solution cool to a maximum of 65° to 70°. Keep a cup of vinegar nearby for rinsing in case the lye solution touches your skin.

Cover the olives with the lye solution early in the morning. Place a towel or cloth over them, and push it down tightly to keep the olives submerged. Stir the olives every two or three hours, until the lye reaches the pits (10 to 12 hours). Judge the amount of penetration by cutting sample olives to the pit with a sharp knife at one or two hour intervals. Lye solutions discolor the flesh yellow-green.

Frequently this lye treatment is insufficient. Some olives neutralize most of the lye and it fails to penetrate to the pits. If the lye has not reached the pits by evening, remove the lye solution and cover the olives with water. Next morning pour off the water and cover the olives with a solution of one ounce (about two level teaspoons) of lye per gallon of water and let stand until the lye reaches the pits completely (may take as long as 30 hours, if the fruit is very green).

Remove the lye solution and discard it. Rinse the olives twice with cold water and cover with cold water. Change the water four times a day, until you can no longer taste lye in the olives (may take as long as seven or eight days). Expose the olives to the air as little as possible during the lye treatment or washing.

Prepare a salt brine containing four ounces salt (about 6½ level tablespoons) per gallon of water. Dissolve the salt thoroughly in water and cover the olives with the solution. After standing for two days, the olives will be ready for use. Store them in the refrigerator.

GREEK OLIVES

Prepare Greek-style olives from mature olives which are dark red to black. Mission olives are commonly used, but any variety will do. Use smaller olives, as the larger ones become soft and shriveled when they are salt-cured. Greek olives are salty and slightly bitter; you may have to acquire a taste for them.

Cover the bottom of a wooden box with burlap. Weigh out one pound of salt for each two pounds of olives. Mix the salt and olives vigorously in the box to prevent mold from developing. Pour an inch-thick layer of salt over the olives.

After one week, mix the olives by pouring the olives and salt into another box, and back into the first box. Repeat this mixing process once every three days, until the olives are cured and edible (30 to 35 days). When the olives are fully cured, sift out most of the salt through a screen. Dip the olives momentarily in boiling water. Drain and allow to dry overnight.

Add one pound of salt to each ten pounds of olives, and mix. Keep the olives in a cool place. Use within one month, or store in a refrigerator or freezer. Just before using, coat the olives with olive oil. (Do not use oil if you plan to use the olives for cooking.) Put the olives in a large pan or box and sprinkle a little olive oil over them. Work the olives with your hands to coat them with oil.

SICILIAN OLIVES

Sicilian-style olives are fermented. Use green olives of any variety, and place them in a barrel or glass-topped fruit jar. Add a level tablespoon of dill pickle spices per quart. Add ½ teaspoon fennel seed per quart, or add a sprig of fresh fennel or dill. If you desire a hotter seasoning, add whole peppercorns and whole red peppers. Garlic is another common seasoning.

Prepare a salt solution of one cup of salt per gallon of water. Add one pint of vinegar to each ten pints of solution and fill the jar or barrel. Store the olives at about 70°. Place a lid on the container, but do not seal. Replace any lost brine. When all gas formation ceases (within about two months) seal and store until the olives have the desired flavor. The olives will remain somewhat bitter and will acquire a flavor somewhat like Spanish green olives (will be different, if spiced).

LYE ACCIDENTS

If you receive lye burns while processing olives, apply vinegar, lemon, orange, or other acid fruit juice, until the lye neutralizes.

If the lye solution has been swallowed, call a doctor immediately. Do not induce vomiting. Administer egg white or milk by mouth. Place the affected person in a reclining position, head lower than the body, and wrap the person in blankets to combat shock.

If lye has splashed into the eye, flush with a stream of running water, bathe the eye(s) with a boric acid solution, and call a doctor.

SUGGESTED READING

Collingwood and Brush. *Knowing Your Trees.* Washington, D.C.: The American Forestry Association, 1964. A good booklet covering 51 common United States trees, with over 260 illustrations and photos showing typical leaves, fruit, bark, flowers. Maps of locations are also given.

Hill, Lewis. *Pruning Simplified.* Emmaus, PA: Rodale Press, 1979. Instructions for all types of pruning, especially for urban environments. How to prune for both appearance and nut/fruit production.

Nature Study Guild Publishers, Box 972, Berkeley, CA 94701. *Desert Tree Finder, Flower Finder, Pacific Coast Tree Finder, Master Tree Finder, Winter Tree Finder, Eastern Tree Finder, Rocky Mountain Tree Finder.*

University of California, Cooperative Extension, Berkeley, CA 94720. *Home Pickling Of Olives* (pamphlet).

Wittrock, Gustave. *The Pruning Book.* Emmaus, PA: Rodale Press, 1976. A Complete summary of all aspects of pruning.

MUSHROOMS

If you ride a bike or walk to school or work, you will inevitably notice mushrooms on lawns, in shady areas, under trees and in open areas. Mushrooms are a source of joy, wonder, superstition, confusion, fear and even disgust. Much of the superstition and fear of mushrooms stems from their sudden and almost magical appearance in an age when they were not fully understood. That some are deadly and others induce visions does not improve the mushroom's reputation. In the minds of many, the phallic shape of the mushroom makes it a kin to the Devil, witchcraft, snakes and toads.

We have just begun to understand what part mushrooms play in the ecology of the earth. The mushrooms you see around town are the reproductive part of the plant, producing spores for the coming generation of mushrooms. Visible mushrooms are produced by mycelium, a white cobwebby material ("spawn" to growers) found a few inches below the soil surface. Mycelium spreads underground, decomposing dead and dying organic material, breaking it down into soil fertilizer for the next generation of plants. Mushrooms are one of the many recyclers of planet Earth.

Everyone invariably asks, "How can you tell if it's edible or poisonous?" You may have heard of one or more methods to determine whether a mushroom is edible or poisonous: checking if an animal has eaten it, seeing if the cap peals, or if the mushroom turns silver-black when boiled with silver. All these methods are worthless in determining edibility.

No quick ways exist to determine edibility. If you intend to gather mushrooms for food, *you must know positively that the specific mushroom you have is edible.* This rule applies to using any wild vegetation for food.

How do you identify specific plants positively? Read books on the subject. Become familiar with the parts of a mushroom and the different types of mushrooms. Check your local college or arboretum for people who know about mushrooms. Mushroom societies give the novice help in understanding the mushroom kingdom. Write to The North American Mycological Association, 3 Ginger Hill Lane, Toledo, Ohio 43623 for the address of the mushroom society nearest you.

While I have enjoyed many mushroom hunts, silently stalking through the forest in search of the Steinpiltz, chanterelles or the wood blewits, I would be irresponsible if I tried, through the limited scope of this book, to teach mushroom identification. If you have a real desire to learn about mushrooms, connect with all possible learning aids.

During a cross-country-and-back bicycle tour in the summer of 1974, my friend Drew Devereux wrote me a letter about his use of wild mushrooms. Drew was the art editor of *The Mycena News*, the San Francisco Mycological Society Newsletter.

"Not long ago," Drew wrote, "I was at a friend's house who is an enthusiastic forager of wild delicacies from the bountiful supply of Mother Nature. Having visions of pizza but low on funds, we decided to create one of the most super pizzas ever to be seen, with mushrooms being the star ingredient. We promptly set off for the local cemetery, where we soon found a large circle of Fairy Ring mushrooms *(Marasmius oreades)* adjacent to one exceptionally beautiful tombstone. While collecting about a pound of these, my friend saw white dots in the distance, which proved to be meadow mushrooms *(Agaricus campestris)* in abundance. We helped ourselves to those and some large puffballs *(Calvatia gigantica)*, one of which was football-sized.

"On our way back to the kitchen, we picked lamb's quarters *(Chenopodium album)* and wild onions *(Allium)* which added a welcome green. From the backyard we took cherry tomatoes from a plant gone wild and some ripe black nightshade berries *(Solanum douglasii)* to insure success to our already promising pizza. Now with a little flour, cheese, oil, and a few spices, we went to work and were rewarded with a delicious pizza. We ate with the satisfying feeling of having gathered all the main ingredients from the wild, without touching any of those germ and bacteria-coated coins that were jingling around in our pockets.

"This mellow scenario can be reenacted just about anywhere during most of the year, with an allowance in the ingredients for what is in season. All it takes is a little patience, practice, and botanical knowledge, plus a lot of curiosity, respect and love for the varied plant life that silently thrives outside your door.

"As mushroom hunters, we are continually opening our eyes to these bizarre and diverse forms of life, while reaping the rewards of our study by seeing, smelling, and tasting some of the finest plants available.

"Identifying a plant or mushroom can become as natural as recognizing a friend on the street or a cauliflower in the produce market. It will turn every walk or hike into a creative learning experience with something new around each corner, making you more in tune with the earth you live on. This awareness and appreciation will reward you in more ways than one."

SUGGESTED READING

Bigelow, Howard E. *Mushroom Pocket Field Guide.* New York: Macmillan Publishing Company, Inc., 1974.

Department of Agriculture, Information Division, Ottawa, Canada. *Mushroom Hunting for Beginners*, publication 861.

Genders, Roy. *Mushroom Growing for Everyone.* Levittown, NY: Transatlantic Arts, Inc., 1970.

Guild, Ben. *The Alaskan Mushroom Hunter's Guide.* Anchorage, AK: Alaska Northwest Publishing Company, 1977. Designed for a specific audience, this book is still useful over most of the United States.

Kavaler, Lucy. *Mushrooms, Molds, and Miracles.* New York: John Day Company, 1965.

Krieger, Louis C. C. *The Mushroom Handbook.* New York: Dover Publications, Inc., 1967. A classic manual of mushroom identification. Originally published in 1935, this well organized manual has 158 illustrations and 32 color drawings. Highly recommended.

Lange and Hora. *A Guide to Mushrooms and Toadstools.* New York: E. P. Dutton, 1972. A very good reference for both the beginning and expert mycologist. The book is full of color drawings and a key helps you find the genus at hand.

McIlvaine, Charles; MacAdam, Robert K. *One Thousand American Fungi.* New York: Stephens Publishing Company, 1973.

McKenny, Margaret. *The Savory Wild Mushroom.* Seattle: University of Washington Press, 1971. A good book for recipes, but don't rely on it for idenfitication.

Miller, Orson K., Jr. *Mushrooms of North America.* New York: E. P. Dutton and Company, 1972.

Mycological Society of San Francisco, Inc., P.O. Box 904, San Francisco, CA 94101. *Kitchen Magic with Mushrooms*, 1973.

National Geographic, October, 1965. An excellent story about mushrooms with color photographs.

North American Mycological Association, 3 Ginger Hill Lane, Toledo, OH 43623. Write for the mushroom society nearest you.

Rolfe, R. T. and E. W. *The Romance and Fungus of the World.* New York: Dover Publications, 1974.

Sanford, Jeremy. *In Search of the Magic Mushroom*. New York: Clarkson and Potter, Inc., 1973.

Smith, Alexander. *Field Guide to Western Mushrooms*. Ann Arbor: University of Michigan Press, 1975. This book refers mainly to the Western United States. Good-quality photographs. Highly recommended.

―――. *The Mushroom Hunter's Field Guide*. Ann Arbor: University of Michigan Press, 1958. This field volume is very popular among mushroom hunters. Includes both black and white and color photographs.

PLANT THE EARTH

The quality of our environment is becoming rapidly and progressively degenerate, but we need not be apathetic and uncaring. One person can do alot. Too many of us toss food wrappers on an already littered roadside. Hikers too often break branches senselessly, roll or throw rocks, scar hillsides or leave a sloppy camp behind. What differences can minor changes make when everything is so bad already? Nothing changes or improves unless individuals vow to change their personal habits. Know with pride that you are not contributing to environmental deterioration. Not only can you refrain from personally adding to environmental deterioration, but you can also teach others by example. Seek ways to improve specific areas, and make sure you leave an area clean before you move on. Don't leave your mess for the sanitation department or someone else.

Johnny Appleseed or John Chapman, a nurseryman, herb doctor, military hero and religious enthusiast of the early 1800s, became a legendary figure in American history, because he fervently distributed apple seeds and sprouts as he wandered through Ohio. The plants he established in the wilds made welcome treats for later wilderness travelers. When Johnny Appleseed left the earth, he left it a better place.

Become modern Johnny and Judy Appleseeds, planting seeds wherever you go. Carry small seed packets to distribute seeds wherever you can, in areas you pass daily. Hikers can introduce favorite food plants along trails and streams. You may think planting seeds is a small contribution, but we've got to start somewhere, making the earth a garden again.

What are the most feasible plants for us to spread? For starters, plant onion, garlic and chives; they commonly grow wild in sandy river banks. Onions and garlic are healthful foods, antiseptics that protect the body from contagious and infectious disease. Mashed raw garlic and onion leaves relieve external bites and wounds. Fresh onions can make a meal come alive. When you use wild onions, pick and use the greens only. If you leave the bulbs, they will continue to multiply and reproduce, year after year.

Tomatoes do surprisingly well in the wild and need less water than you might think. Once the seeds have sprouted and the plants establish themselves from spring rains, tomato plants will do well on their own. I have seen tomato plants in shaded canyon areas produce tomatoes for at least three years. When you harvest wild tomatoes, squeeze out some seeds, so more plants will sprout up. Tomato leaves and stems make a good dye that requires no mordant. They dye wool and silk reddish brown to light yellow, depending on the season.

Scatter fennel and celery seeds, especially in moist environments. Once celery is established, you'll have a permanent supply of this good salad addition. Fennel, also called wild licorice, is able to endure drier environments, such as vacant lots and roadsides. Eat the tender bases of fennel stalks as well as the celerylike bases of the leaves. The abundant fennel seeds make good licorice-tasting tea which has a great variety of medicinal uses.

Jerusalem artichokes are a potatolike tuber of the sunflower family. Like the potato, each "eye" will produce a plant. Moist, loamier soils are best for this plant, so the bulbs can expand easily. Planted in winter or early spring, the tubers will be ready to harvest the following autumn, after the plant has flowered and died. Take the biggest ones and rebury the rest for the next season. Eat Jerusalem artichokes raw in salad, boiled, sliced and fried, or boiled and whipped into "mashed potatoes."

Distribute the seeds of kale, collards, Swiss chard and broccoli. Although wild greens tend to have smaller leaves, they provide a semipermanent source of vitamin-rich food.

Plant radishes, carrots and beets in the "wild," and harvest the roots later. If the root is not used the first year, all but the outer layer becomes too fibrous to eat. The edible greens of beets and radishes will be available even after the roots are tough. Wild carrot seeds can be harvested, winnowed and used for seasoning.

For the more devout modern Appleseeds, try planting sprouted avocado seeds in empty lots or along roadsides. Grape vine and young fig trees will also do well. We really can change the world—but not until we change ourselves.

IV
THE PROBLEM WITH POLLUTION

Pollution is a fact of city existence. Running from pollution, to where the air is fresh and clean, is not a choice most people have. We are city dwellers, and we must do the best we can to help initiate programs to clean up the environment.

We should not accept dirty air. Not only should we take personal action to protect our bodies from pollution, we should also do our part to produce less pollution, and to seek out ways to fight pollution in our community.

ALUMINUM POISONING

In "advanced societies" daily ingestion of aluminum particles is virtually inescapable. Since this time yesterday, millions of us have ingested "pure foods," never suspecting we were consuming compounds of aluminum from aluminum utensils, aluminum foil, alum baking powders, drinking water and many foods.

Aluminum foil is, quite literally, all around us—used for wrapping and packaging chocolates, cigarettes and cheese. How many millions of people instinctively reach for that roll of aluminum foil to wrap foods before returning them to the refrigerator or freezer? Most dry foods bought in grocery stores today are packaged in aluminum or aluminized packages. Brewers make wide use of aluminum in fermentation vessels. Refrigerators now have aluminum drawers and ice cube trays. Many soft drinks and beer cans are aluminum. The dairy industry uses aluminum for transporting milk in large tankers. Chemists, physicians, public health officials, food manufacturers, morticians and dye manufacturers all use aluminum, despite the fact that serious investigations into aluminum poisoning began many years ago.

Aluminum is probably the chief poison in our food. The average United States resident unknowingly suffers more from aluminum toxicity than from DDT poisoning. Aluminum reacts to acids and alkalies: almost all foods cooked in aluminum produce some kind of aluminum compound. Eggs produce aluminum sulfate; salted meat and vegetables produce aluminum chloride. The list goes on and on. (Some vegetables naturally contain salt, and nearly all canned and frozen vegetables are presalted. Aluminum chloride will be formed even when no salt is added to the cooking water in the aluminum vessel.) Temperature and length of cooking time affects the amount of aluminum dissolved in foods or beverages.

SCIENTIFIC RESEARCH

In the late 1860s and early 1870s, 33 pages were devoted to aluminum poisoning in *Allen's Encyclopedia of Materia Medica*. This book was published before aluminum was used in cooking utensils or in baking powder.

Between 1881 and 1887 a rapid increase in gastritis, appendicitis and ulcerative diseases, both in animals and humans, was noted in Europe, England and America.

By 1892, German scientists discovered that aluminum was not suitable for storing alcoholic beverages.

Between 1911 and 1920, aluminum cookware began to be produced and used in the United States. A multimillion-dollar industry was born.

An American doctor, C. T. Betts, published his deductions about the potential harm of aluminumware in 1913, and his fight against the use of aluminum began. Dr. Betts found that aluminum was blended with soda and sulphuric acid in baking powder to form the gas that makes dough rise. He hypothesized that the aluminum in cooking utensils dissolves in water, and when combined with the stomach's powerful acids, forms gas in the gastrointestinal tract.

In 1920 the United States Federal Trade Commission began a five-year survey to determine if illness or death resulted from the ingestion of aluminum compounds. The report was suppressed.

Detailed reports made by Dr. Leo Spira of London showed that patients suffering with aluminum poisoning symptoms were relieved of all symptoms when aluminum compounds were eliminated from their diet (including food prepared in aluminum and water in which aluminum might be present).

Dr. Eastes, in 1930, tested aluminum cookware with startling results. Milk heated for 20 minutes in an aluminum vessel detached 140 grains of material per gallon, mostly aluminum. Gooseberries stewed in an aluminum saucepan detached 31 to 85 grains per gallon from the pan. The juice of 3 lemons to 500 cc. of water, detached 25.6 grains per gallon.

From 1932 to 1942 Dr. R. M. Le Hunte Cooper published reports of his findings regarding food contamination by aluminum. Dr. Cooper was one of the first homeopaths to call attention to the poisoning effects of aluminum in humans.

By 1936, cancer had climbed rapidly toward becoming one of the top five killers. The rise in cancer occurred at the same time that aluminum compounds were introduced throughout the world. Cancer incidence increased from 173 per million people in 1842 to 1563 per million in 1936.

In 1952, magazine publisher J. I. Rodale ran a series of articles summarizing data he had been collecting over many years, showing the undesirability of aluminum cookware. Rodale described the use of aluminumware as "one of the great blunders of our time along with the use of chemical fertilizers."

Evidence regarding the ingestion of aluminum, which has piled up during the past seventy years, seems overwhelmingly conclusive.

PHYSIOLOGICAL EFFECTS

Aluminum is not one of the essential minerals in forming body structure or regulating metabolism. Its introduction into the body in any form—food, medicine, drug, or from pots and pans—is treated by the body as foreign matter. It immediately begins to irritate the stomach and alimentary tract. If aluminum entered the body in large particles, evidence shows that it would be discharged from the body with little or no damage to the tissues. But the body acids, and the acids and alkalies of foods and liquids, combine to change microscopic aluminum particles into sharp-edged crystalline "salts." These salts cut, slice, gouge and abrade, as they travel through the body. Salts that are small enough to pass through the digestive membranes become a source of obstruction throughout the body's vascular system, causing weakness and enervation.

Aluminum residues produce toxic effects, even in small doses, over a long period of time. Toxic effects (pathological lesions in the stomach and spleen, thrombosis, necrosis, pigmentation and fibrosis) indicate the presence of a poison clumping and breaking down corpuscles. Aluminum toxicity causes duodenal ulcers, gallstones, different types of skin conditions (such as eczema), leg cramps, excessive perspiration, low blood pressure and pyorrhea of the gums. Aluminum directly reduces the body's resistance to infection, by destroying vitamin C.

Aluminum salts rob the body of potash, sulphur and silica, reducing the effectiveness of the glandular system, especially the pituitary gland. This can result in the uncontrolled growth of tumors. Even in infinitesimal amounts, aluminum salts have an adverse effect on the body's metabolism.

Underarm deodorants composed of aluminum compounds cause many neurological disturbances. Aluminum enters the body through the armpit, penetrates the lymph glands, and has easy access to the brain.

Why aren't we all ill from aluminum's effects on our bodies? Some of our bodies are less susceptible than others; they acquire a sort of immunity. Some bodies have greater excretory and/or eliminative capabilities which are short-term advantages. Aluminum may be taken into the system for years, even decades, without noticeable harmful effects; nevertheless, measurable alterations in bodily tissues are taking place at a microscopic level, and are steadily accumulating towards inevitable disease, illness, impairment and dysfunction.

Chronic ailments of many years standing have completely disappeared, shortly after discontinuing aluminum usage. A top-quality protein supplement will combine with heavy metals and precipate aluminum out of the body. Vitamin C will also precipitate out heavy metals; take as much as 5,000 to 10,000 mg. vitamin C per day (Dr. Sam Walters). Also, take a good-quality broad-spectrum mineral tablet with each meal.

HOME RESEARCH

You need not believe someone else's conclusions. Here are some simple tests you can do yourself to determine whether aluminum particles leach into cooking water and food. If you've cooked with aluminum for any length of time, you have probably noticed many of these properties. For best results use thoroughly clean aluminum utensils.

Boil undistilled water in aluminum for half an hour and an equal quantity of water in porcelain for the same length of time. Pour the boiled water into separate glass jars and let stand for an hour. Shake the jars gently with a circular motion. The water boiled in aluminum will be cloudy; this cloudiness is caused by a precipitate called aluminum hydroxide. The water boiled in porcelain will be clear.

Place a quart of water in a clean aluminum pot. Add a teaspoon of salt and one of baking soda. Boil an hour, adding more water as necessary. Remove from the heat and let stand two hours. Pour into a glass container and note the milky appearance caused by the loosened aluminum particles. These particles will settle out in a day or two, but give unmistakable evidence of the solubility of aluminum.

I witnessed the following experiment in the name of science. Because I respect all life forms, I do not recommend anyone actually perform this experiment. Two quarts of water were boiled in an aluminum pan, two quarts in stainless steel and two quarts in porcelain. When the water was thoroughly cooled, three goldfish were placed in each pan. Within six hours the fish in the aluminum pan were dead; the fish in the other two vessels lived five days before they were returned to their aquarium.

Boil either hard or soft water in aluminum and allow it to stand overnight. In the morning when you stir the water with a spoon, a heavy gray cloud will rise. The cloud will be even heavier in hard water.

Boil cherries or grapes in aluminum, and allow them to stand in the pan for twelve hours. You will find little pits or holes in the pan. Where did all the aluminum go?

Cooking tomatoes, applesauce and rhubarb for five minutes in aluminum will brighten the aluminum considerably, and render even the grimiest utensils "clean." Where did all the "dirty" go?

Boil cabbage in aluminum and it will turn the pot dark or black. Boiling butterscotch pie filling for just a few minutes in aluminum turns the pan brown to dark green. Why?

Leave lemonade in aluminum overnight; it becomes bitter and "off" tasting. Cook and cool jello in aluminum, and it will taste bitter.

Cook peeled potatoes in aluminum and they will develop dark streaks. Left in aluminum twelve hours, potatoes will turn yellow; applesauce will turn green.

Tea looks cloudy if you steep it in an aluminum pot.

Perhaps the best use for all aluminum utensils and cookware is as a donation to the Boy Scout metal drive. The best types of cooking utensils are high-fired ceramics, glass, stainless steel or cast iron. Though not inexpensive, they are durable utensils for all culinary purposes.

SUGGESTED READING

Health Research, 70 Lafayette Street, Mokelumne Hill, CA 95245. *The Story of Aluminum Poisoning.* A comprehensive pamphlet describing in detail why aluminum is poisonous. Contains information not found elsewhere. Not available widely; you probably will have to write to the publisher.

Mittleman, Jerry. "That Toxic Aluminum" in *Acres U.S.A.*, October, 1978.

Rodale, J.I. and staff. *The Prevention Method for Better Health.* Emmaus, PA: Rodale Press, 1952. A comprehensive manual of nutrition and food. Includes a section on aluminum poisoning.

GARLIC SAVES

During the Middle Ages, when the bubonic plagues ravaged Europe, it was claimed that those who ate garlic daily were not infected. Eating garlic or wearing it around the neck to prevent or cure colds is scoffed upon today. The preventative and detoxifying properties of garlic are not just folklore. Scientific evidence supports the claim that garlic protects urban dwellers from the ravages of smog and a host of other physical disorders.

LEAD AND MERCURY

Heavy metals in the atmosphere, such as lead and mercury, have been known to poison the blood, destroy the erythrocite membrane (a red corpuscle which contains hemoglobin). Hemoglobin carries oxygen to the lungs and body tissues, and carbon dioxide from the tissue back to the lungs. All city dwellers—bicyclists and runners in particular—are vulnerable to lead poisoning.

Early and low-level symptoms of lead poisoning include vomiting, headaches, fatigue, poor appetite, irritability and depression. Sometimes no apparent symptoms are present. One of the easiest and cheapest solutions to lead poisoning is including garlic in the diet. Because lead is so prevalent in the air we breathe, urban dwellers must protect their bodies from this ubiquitous poison. Garlic is a top-quality prophylactic against lead poisoning.

Garlic will precipitate lead and mercury out of the body. Selenium (a chemical element of the sulfur group) and other sulfur compounds bind to lead and mercury, and garlic contains more selenium than most other vegetables. Most selenium is eliminated when the garlic is cooked or boiled, so raw garlic must be eaten. Eating raw garlic requires some adjusting, both because of taste and social ramifications. If you absolutely cannot eat raw garlic, but want the healthful benefits, buy garlic oil capsules in your health food store or co-op. They come regular and odorless—take your choice.

Almost 30 years ago a German physician, Dr. J. Klosa, reported that garlic oil has remarkable antibiotic powers (*German Medical Monthly*, March 1950). The most notable garlic research has been done in Hiroshima, Japan by Doctors Kitahara and Ikezoe. Until their study was underway, expensive chelation treatments had been the only known effective medical treatment for detoxifying the body of heavy metals. Doctors Kitahara and Ikezoe have proven that raw garlic extract is effective in protecting the body from the toxic effects of heavy metal poisoning.

The body's epithelial cells are greatly damaged by pollution. These cells make up the skin and line the mouth, nose, throat and lungs. Garlic helps maintain a healthy metabolism and growth of epithelial tissue, even in a polluted environment. The destruction of red corpuscles, releasing hemoglobin into the surrounding fluid, is called hemolysis. In 1975 Kitahara and Ikezoe showed that garlic prevents hemolysis.

Using four test tubes of red blood cells, the doctors added lead to the first one, mercury to the second, lead and raw garlic extract to the third, and mercury and raw garlic extract to the fourth. If hemolysis takes place, the clear fluid in the test tubes becomes red to dark red, from the release of the red pigment, hemoglobin. After 24 hours the first two test tubes (lead and mercury without garlic) were bright red, indicating that hemolysis had taken place. Test tubes three and four (into which garlic had been added) showed no hemolytic effect in the presence of heavy metals. Dr. Kitahara's booklet, *Biological Aspects of Kyolic*, includes these test results. (Kyolic is the Japanese name for the raw garlic extract.)

Dr. Kitahara gave two groups of rats mercury. The rats given raw garlic extract eliminated mercury through the fecal matter two to three times faster than the other group. The same results were noted with lead and copper. This information should be of great interest to urban bycyclists and runners. Including garlic in the diet is immediate personal action, protection of the body against the ravages of pollution.

A recent study by researchers at the Brain Bio Center in Princeton, New Jersey showed that Vitamin C and zinc also protect against lead poisoning. During this study, led by research biochemist Rhoda Papaioannou, 22 battery plant workers took two grams (2,000 mg.) of Vitamin C and 60 mg. of zinc daily. Battery plant air was so filled with lead that many workers showed signs of lead poisoning. The level of lead in the workers' blood was measured before the study, then 12 to 24 weeks later. After 24 weeks, the workers' average blood lead level had dropped 26%. The workers showed decreases in lead poisoning. The researchers theorized that the Vitamin C and zinc may have prevented the absorption of lead from the digestive tract, and removed lead already in the system.

CHOLESTEROL

Dr. Alan Tsai, Ph.D. (Michigan School of Health), has been testing rats and humans for garlic's effect on cholesterol levels. He fed two groups of rats high-cholesterol diets; he gave one group garlic and the other none. The rats without garlic showed a 23% increase in cholesterol, while those fed garlic had levels that rose only 4%. Dr. Tsai notes that the incidence of cardiovascular and other diseases is lower in countries whose populations consume large amounts of garlic.

The "Indian Journal of Nutrition and Dietetics" (Vol. 13, no. 1, 1976) reported on the tests of K. K. Sharma and three coworkers in the Department of Pharmacology and Medicine, S. N. Medical College, Agra, India. They fed ten healthy young volunteers high-cholesterol meals. Within four hours, as expected, their cholesterol levels shot up. Those who ate garlic with the meal (raw or cooked) maintained low-cholesterol levels.

Other studies from the "Indian Journal of Nutrition and Dietetics" (Vol. 12, 1975) show that onions (a close relative of garlic) also lower blood cholesterol levels. Including onions in the diet may reduce chances of heart attack and stroke. Dr. A. S. Truswell, professor at the Queen Elizabeth College of London University, fed some subjects high-fat meals with onions and some without onions. When onions were included in the high-fat meal, the fat was neutralized.

OTHER POSITIVE RESULTS

Dr. F. G. Piotrowsky of the University of Geneva showed that garlic lowers high blood pressure by opening up tight blood vessels. It relieves dizziness, angina pains, and headaches (*Praxis*, July 1, 1948).

In 1962, Japanese scientist Fujiwara discovered that allicin in garlic increases the body's capacity to assimilate Vitamin B1 (*Pakistan Medical Times*, May 16, 1961). During the 1965 flu epidemic, Russia purchased 500 tons of garlic. Advertisements in the *Evening Moscow Journal* advised eating more garlic for its prophylactic qualities in preventing flu.

The American Lung Association reported in June, 1978 that eating garlic and onions during the "cold season" will cause an overproduction of mucus in the stomach and respiratory system; clear breathing passages; relieve existing colds, and increase resistance to future colds. Dr. Albert Schweitzer used garlic (with positive results) to treat typhus and cholera.

SUGGESTED READING

Bragg, Paul C. *The Shocking Truth About Water.* Burbank, CA: Health Science, 1972. Covers a complete rundown of how we poison our water and how it affects our health. Must reading for those who believe "things ain't so bad."

Fujiwara, Ph.D. *Pakistan Medical Times*, May 16, 1961. Dr. Fujiwara's report indicates that the allicin in garlic increases the body's capacity to assimilate Vitamin B1.

Grossman, Shelly. *Understanding Ecology*. New York: Grosset and Dunlap, 1970.

Indian Journal of Nutrition and Dietetics, Vol. 13, No. 1, 1976. The tests of K. K. Harman and three coworkers in the Department of Phar-

Peace/Urban Wilderness   72/250   hfK

Khan, M. A.; Bederka, John, eds. *Survival In Toxic Environments*. New York: Academic Press, 1974.

Kitahara, S., Ph.D.; Ikezoe, Ph.D. *Biological Aspects of Kyolic*. Dr. Kitahara's and Dr. Ikezoe's report on test results proving that garlic prevents hemolysis. If you are a doctor or involved in research, you can request a copy of this report by writing to Dr. S. Kitahara, Wakunaga Clinical Research Center, 1624 Kodachi, Takada, Hiroshima, Japan.

Klosa, J., Ph.D. *German Medical Monthly*, March, 1950. Report on the antibiotic powers of garlic.

Piotrowsky, F. G., Ph.D. *Praxis*, July 1, 1948. Dr. Piotrowsky of the University of Geneva reports that garlic opens up tight blood vessels and lowers high blood pressure.

Rain, Journal of Appropriate Technology. 2270 N.W. Irving, Portland, Oregon 97210.

Robinson, John. *Highways and the Environment*. New York: McGraw-Hill, 1971.

Truswell, A. S., Ph.D. *Indian Journal of Nutrition and Dietetics*, Vol. 12, 1975. Report on the value of onions.

Watanabe, Tadashi. *Garlic Therapy*. New York: Japan Publications Trading Company, Inc., 1974.

Wickenden, Leonard. *Our Daily Poison*. New York: Devin-Adair Company, 1956. A shocking 186 pages revealing how DDT, flourides, hormones and other chemicals affect modern people.

Wise, William. *Killer Smog*. New York: Ballantine Books, 1970.

INSECTICIDES

We need not subject ourselves to the harmful fumes of insect poisons. Herbs and other substances provide viable alternatives to eradicating the "pesty" problem of insect invasion.

When summer heat produces a proliferation of ants, many of us go to great lengths attempting to eradicate these tiny creatures. Ant poisons are not always effective, and most are potentially harmful to us and our pets.

Let us review how ants operate and why they are a "problem." When an ant colony exhausts its food supply, a "find food" command goes out to the scouts. Ant commands are relayed by chemicals secreted onto the ants' lower bodies. To receive a message, an ant merely touches the secreted chemicals with its antennae. When locating food, ant scouts retrace their own chemical trail back to the nest and secrete the message "food." Troops leave immediately to bring back all the food the scouts cannot carry. The returning troops bump the emerging troops to confirm "food ahead." When the food is gone, the returning ants report "no more food" by means of a new chemical, and the outgoing ants return.

Solving ant problems involves confusing and redirecting their communication system, providing them with a chemical communication we want them to believe. Either apply a chemical that neutralizes the acid (their base chemical transmission medium) or apply chemicals to their antennae that they interpret as anathema. Apply a sufficient amount of the chemical so that the nest gets the message, "It's horrible out there."

As many of us discover, killing off the visible ants on the kitchen counter is ineffective. The nest continues to communicate "food out there" and will continue to send troops out. The ideal moment for eradication is when you first find scouts searching for food. They must be convinced they have entered a veritable Ant Hell.

The presence of ants in your home is an indication of carelessness.

Ants are in no way bad. These wonderful little creatures help us refine our actions by relentlessly pointing out all those little details we neglect: the crumbs on the floor, the syrup on the stove, etc. Ants are not dirty—we are dirty and careless. The ants are cleaning up our mess!!

The industriousness of the ant was used in the Bible as an example to the lazy. "Go to the ant, thou sluggard; consider her ways, and be wise: Which having no guide, overseer, or ruler, provideth her meat in the summer, and gathereth her food in the harvest." *Proverbs 6:6-8*

White Tower, Inc. Research Division, Los Angeles, spent over a month experimenting with the common household chemicals available at the local market. Nearly three dozen combinations of liquids and spices were tried. Cloves were the only single "natural" item that caused every ant (even a fast moving thick column) to recoil. When the antennae touched cloves, they would turn and flee. Sprinkling powdered cloves is impractical since many ant invasions take place, wholly or in part, on vertical surfaces.

The ants recoiled from Basic H liquid detergent, but slowly made their way through it, when it began to dry. Cloves mixed with Basic H and a little water proved to be the best combination. This mixture offers an ant repellent/barrier that lasts a long time before drying, clings to vertical surfaces and is inexpensive. Coarsely granulated cloves work best. If whole cloves are used, cut them in half (to expose the germ), and cook in water at 130° for 10 minutes (don't go over 130°). Cool and use the water and the cloves. Straining is not necessary, but a spray bottle with a filter is essential. A mix of approximately 25% Basic H and 75% clove water works well. Cloves are available in the local supermarket spice section, and any health food store or coop. Basic H is available from all Shaklee distributors.

Once the ants plunge their antennae into the mixture, they stop for a long time to clean their antennae. A light spray coats the ant's body and antennae, seeming to nullify or disrupt their ability to detect other chemicals. A light mist of this mixture will turn back a roaring column of ants. You need not kill ants. Lightly covering their bodies will give them a chemical "message."

FLEAS

California Indians would scatter the strongly aromatic leaves of the California or laurel bay *(Umbellularia californica)* whenever fleas were prevalent. Try scattering these bay leaves in the dog house. Soak the dog in a bath of bay water and use a fine comb afterwards. Scatter California bay leaves in cupboards or place them in flour containers to help repel insects. Young eucalyptus seed pods strung together make a natural flea collar for the dog. Rubbing dried fennel leaves *(Foeniculum vulgare)* into the dog's hair is also effective. Or wash the dog with a strong fennel leaf infusion and comb out the dog's hair.

MOSQUITOES

During summer months, we are often plagued with mosquitoes, biting flies and other winged insects. Insects can be a nuisance, unless you know how to deal with them. Most people are not interested in the fact that only the female mosquito bites (in order to obtain protein for her eggs). Nor do those bitten by mosquitoes care if mosquito larvae are a part of the food chain of certain fish. Most people are interested only in knowing how to keep mosquitoes away.

Most mosquito repellents on the market are of dubious quality; effectiveness is often variable from day to day, and from person to person. Certain repellents have a good reputation: Amway's D-15 Insect Repellent, Cutter's Insect Repellent and Sportsmate II. Cutter's and Sportsmate are available as creams, and are neither sticky nor greasy. Cutter's is also available in spray and stick, and Amway's as spray only.

The main active ingredient in effective repellents is *N-Diethyl-meta-toluamide*. It is found in concentrations from 14% to 35% in commercial products. If you have the persistence to rummage through surplus stores, military insect repellent can be purchased which contains approximately 70% of this active ingredient. A happy bonus is that it is far cheaper than its commercial competitors.

If you find yourself in an insect-infested area with no commercial repellent, common plants can be used to keep away biting, blood-sucking insects.

Laurel sumac *(Rhus laurnia)* is an evergreen shrub, common in the chaparral of West Coast regions. Crush the leaves by rubbing them between your hands; your hands will be oily and sticky with natural repellent. Rub your hands over the exposed parts of the body. Mosquitoes in particular are repelled by the strong aroma of this oil, and will seldom land, even though they may buzz closely around you.

Parsley *(Petroselinum crispum)* is the herb garnish most of us leave after eating the meat and potatoes. Fresh parsley leaves, crushed and rubbed over the exposed parts of your body, are an effective insect repellent.

Aloe vera, (a succulent member of the lily family) has a thick juice which, when applied directly to the skin, repels mosquitoes.

Pennyroyal *(Hedeoma pulegiodes)* is available mostly in the Eastern states. The fresh leaves can be crushed and rubbed on the skin. The strong perfumy smell keeps mosquitoes away.

Plaintain *(Plantago sp.)* leaves, though not effective as a repellent, will relieve pain. Crush the leaves and apply to bites. Alma R.

Hutchens, in her excellent book *Indian Herbology of North America*, states that "the juice of the [plantain] leaves will counteract the bite of poisonous insects. Take one tablespoon every hour, at the same time applying the bruised leaves to the wounds."

It seems that people with thiamin deficiencies are more susceptible to mosquito bites. Many people have reported that 300 mg. of thiamin (vitamin B1), taken daily before and during an outdoor adventure, provides relative immunity to mosquito bites. Try increasing the amount of natural apple cider vinegar in your diet during the summer months. Vinegar will keep mosquitoes from biting.

NOISE POLLUTION

We are continually bombarded with sound waves from both the air and the ground. American Indians heard horses from miles away by putting their ears to the ground, and *feeling* the sound vibrating the earth. Noise travels through the ground today also. Every time the earth is pounded, exploded, beat, cut or graded, the vibrations travel, often great distances through the earth.

How can we survive the effects upon mind and body of noise pollution in the atomic, jet age? People were able to tune out noise in the 1930s and 1940s, but now we live in another era with a wholly different environment. Noise is everywhere and cannot be escaped. We work in loud and noisy factories, drive on asphalt and concrete roadways, ride bicycles down city streets, hear jets and helicopters overhead, and listen to emergency sirens. These sounds have direct effects upon our bodies; they dilate our veins and arteries and eventually give us high blood pressure.

NOISE, STRESS AND DISEASE

Living near a major airport, constantly hearing the roar of jet engines, may substantially reduce a person's lifespan, according to a study released in 1978 by UCLA professor William C. Meecham. The UCLA study compares a neighborhood bordering on the Los Angeles International Airport (people live directly under landing approaches of several hundred jets daily) with a neighborhood just six miles away.

During 1970-1971, the high noise area had 20% more deaths. The occurrence of virtually every disease was higher in the high noise area, from the fear, stress and anxiety associated with the tremendous noise of the jets overhead. Fatal strokes and other cardiovascular problems were 39% higher, and deaths from cirrhosis of the liver were 140% higher. Meecham speculated that because cirrhosis of the liver is alcohol related, alcoholism is also higher in high noise areas.

This study has obviously been challenged, but according to Meecham: "The details we may argue about for a decade. But the fact of the matter is a reasonable person only has to stand under those planes to realize damage is being done."

Other studies have linked birth defects and nervous breakdowns with high noise levels. Dr. Colette L. Cunningham of Newport, Rhode Island has stated that stress can trigger diabetes, hypertension, alcoholism, drug addiction and cancer.

If noise causes stress, and stress is associated with cancer, it may not be long before the nation's doctors are openly declaring that noise is a direct cause of cancer. Our thinking has a great deal to do with our physical state of being. Cancer could represent a strong self-destructive urge, and is a more socially acceptable way to die than suicide. Psychologist Dwight Bulkly claims to have over 2,000 case histories to back up his theories that there are subconscious reasons for all our illnesses and troubles. He states that after specific incidents of stress, self-determined distortions of the bodies' cells, tissues, and functions follow in a precise manner.

INSULATION

A few remedies are available to individuals independent of government action (or inaction).

Hidden sources of noise are very difficult to counteract. Motor vehicles travelling on asphalt produce subsonic noise (cannot be heard) that enters the human body through the ears, skin and bones.

The most obvious personal solution is to wear top-quality acoustically designed earplugs. Unless you are among the fortunate few who drive a sound-deadened car, you should definitely wear earplugs during asphalt and concrete roadway travel.

Earplug wearing is especially critical when operating or riding in motor vehicles with loud engines, a loud muffler, and/or poorly designed vehicles that allow noise to proliferate. Motorcycle and bicycle riders would benefit by wearing earplugs in extremely noisy downtown areas.

Some earplugs produce an annoying "whistle," causing more noise irritation than surrounding noises. If you plan to wear earplugs while riding a motorcycle or bicycle, be wary of purchasing earplugs with a hollow-end design (the earplug end has a hollow cavity). Noise on a motorcycle is usually unbearable. If you are stuck with the wrong kind of plugs and can't afford another pair, pull your 100% wool hat down over your ears to avoid the whistling or wear a helmet.

I am sad to report that walking barefoot, especially in a city environment, can be harmful to the body. Sound waves enter our bodies through feet and leg bones. The best insulation against sound waves is sturdily-constructed, rubber-soled shoes and thick, 100% wool socks. The original Kelso Earth Shoes (now defunct) were truly miraculous shoes; perhaps some of the imitators are equally good. Rubber-soled, leather or canvas-topped hiking shoes (such as the Austrian Kletterschuh), or low-cut boots are your next best defense. Synthetic and leather soles and heels transmit exposed vibrations.

Top-quality wool or corduroy hampers noise attacks. Knowledgeable yogis do their asanas on wool mats, because wool absorbs sound waves from the earth and lessens the effects of sound on the body. Good quality wool is not hot and scratchy, but extremely comfortable. Be careful to cover your feet and head. The bottom of the feet have the most nerve endings in the body.

For noise survival in the home, surround both your house and the perimeter of your property with an insulating layer of thick hedges, bushes, bushy trees and tall leafy plants. A layer of vegetation offers protection from the continuous bombardment of subsonic sound waves, and provides an abundant oxygen supply to counteract the ravages of smog.

SUGGESTED READING

Anthrop, Donald. *Noise Pollution.* Lexington, MA: Lexington Books, 1973.

Beranek, Leo L. *Noise Reduction.* New York: McGraw-Hill Book Company, 1960.

Bragdon, Clifford, ed. *Noise Pollution: A Guide to Information Sources.* Detroit: Gale Research, 1977.

Carmen, Richard. *Our Endangered Hearing.* Emmaus, PA: Rodale Press, 1977.

Scheibel, M. Barbara. *Noise: The Unseen Enemy.* New Haven, CT: Pendulum Press, Inc., 1972.

MIND POLLUTION

Television can be mind polluting or mind expanding: the choice is up to you. Here is another area where the individual can have immediate control over lifestyle and environment. Don't let television control you; you can control your television viewing to reap the greatest benefits. Television has the potential of revolutionizing communications and education. Learn to use this tool; don't let television control you.

To advertisers, television is a series of sales pitches sandwiched in between hopefully alluring programs. The content of the programs is insignificant, as long as it compels large numbers of viewers to sit still long enough to watch the commercials. Unfortunately, producers and advertisers are more concerned with selling than meaning. Too often they sublimate quality for a "package" they hope will be watched by a substantial segment of an amorphous entity called "the television-viewing audience." Only if they attract a substantial audience can they attract advertisers.

Complaining that TV shows have degenerated is passive. Write to a TV station and tell them how you feel. You and I can change the world, individually and collectively. During the 1950s, a young man who loved Laurel and Hardy films was shocked to see them removed from the air. He got two friends to help him write 500 postcards and letters requesting the series be reinstated. Within a month, whether by coincidence or as a result of this man's efforts, Laurel and Hardy were back with us every night.

Sharing complaints without action will not change a thing. Remember that you control the dial. Watch only shows that are uplifting and helpful. Listen actively, perhaps even take notes, and review them to map your perceptions and changes in perceptions. Keep a dictionary nearby; do not assume you know the meaning of each new word as it sails past you. When an intriguing topic presents itself, begin research on your own. Following up on what you learn will enrich your life and sharpen your senses.

Begin regarding TV as a gigantic neutral tool. It can open your eyes to new learning or blind you with dullness and passivity. The choice is yours—the responsibility is yours. You control the dial.

WHO'S IN CONTROL TEST

1. How long do you sit in front of the TV each day?
   A. ( ) Not at all   B. ( ) A short time
   C. ( ) Quite a while   D. ( ) A long time

2. What are your mind and body doing while you sit watching TV?
   A. ( ) Ceaseless directed activity   B. ( ) Considerable directed activity   C. ( ) Occasional directed activity   D. ( ) Hardly anything

3. How much of yesterday's TV viewing do you remember?
   A. ( ) Almost all   B. ( ) A good deal   C. ( ) Some
   D. ( ) Little, if any

4. How much of your TV viewing has directly benefitted your (or your family's) life?
   A. ( ) Almost all   B. ( ) A good deal   C. ( ) Some
   D. ( ) Little, if any

5. What portion of your regular choice of programs is goal directed?
   A. ( ) All of it   B. ( ) A good deal of it
   C. ( ) Some of it   D. ( ) Little, if any

6. What portion of the programs you watch is inane comedy, violence or sex?
   A. ( ) Little, if any   B. ( ) Some
   C. ( ) A good portion   D. ( ) Almost all

7. Do your choices generally require personal cooperation, commitment, physical involvement or active thinking?
   A. ( ) Always   B. ( ) Often   C. ( ) Sometimes
   D. ( ) Rarely

8. How often do you let the TV set play "just for the sound" with no one actively watching it?
   A. ( ) Never   B. ( ) Rarely   C. ( ) Sometimes
   D. ( ) Often

9. How often do you end a day's TV viewing feeling bored, dissatisfied, frustrated (even angry), groggy or mindlessly "neutral"?

   A. ( ) Never   B. ( ) Occasionally   C. ( ) Frequently
   D. ( ) Most of the time

10. How often have you wanted to see "something good," but settled for whatever was on?

    A. ( ) Never   B. ( ) Occasionally   C. ( ) Frequently
    D. ( ) Most of the time

11. How often do you choose to watch TV rather than read, exercise or involve yourself in community activities?

    A. ( ) Rarely   B. ( ) Occasionally   C. ( ) Frequently
    D. ( ) Regularly

12. How often do you turn on the TV, knowing that you will probably fall asleep in front of it?

    A. ( ) Never   B. ( ) Occasionally   C. ( ) Frequently
    D. ( ) Most of the time

Scoring: A = 4 points; B = 3; C = 2; D = 1. Add up your total points. High score, 40–48 points, means you are in control; a medium score, 25–40, means read on, you are not in full control; a low score, below 25, means the TV controls you.

After honestly reviewing this test, you will be able to see if you are in control of the TV, using it wisely as the marvelous tool it can be; or if the TV "image-makers" are controlling you, keeping you in a near-hypnotic trance, day after day.

Those who rate high on the TV test may be exercising, taking notes, ironing, mending, cooking and doing busywork while watching TV (Question 2). They remember important highlights (Question 3), which they apply toward their goals (Question 5). High scorers generally feel that TV directly benefits their lives (Question 4). Their viewing choices tend to be instructional or factual, requiring mental involvement and active, directed thinking (Question 7). High raters seldom settle for "whatever is on" (Question 11); they choose their viewing selectively; they end the evening feeling as if they have learned something (Question 10). If you rate high on the test, you may have little need for the following information; you have apparently already learned and applied it.

If you rated low on the test (you know who you are), you will benefit immensely by taking a closer look at TV.

Medium-low scorers need to become more aware of the mind.

Regardless of your age or educational background, do not allow your mind to stagnate. Recognize that TV is a powerful mind conditioner, even a "brainwasher," and, as such, must be used cautiously. Guard against basing viewing choices on: "makes me laugh," "cute," "exciting," "sad," "lively," or just "entertaining." Look for information, plans of action and challenges.

Remember: Your nonwork hours are just as valuable as your working hours. Do not allow your brain to be assaulted with shallow TV offerings, any more than you allow people to wander in off the street and chatter at you while you are at work.

SELECTIVE VIEWING

Television is like an unexplored land. If you have not yet dared to begin the exploration, take immediate steps to change the situation. If you watch TV every evening, hoping to find something good on the tube, begin planning the week's viewing ahead of time. How does one determine what to watch and, perhaps more importantly, how to watch?

TV guides are the Rosetta Stone of the language of television. Subscribe to as many as you can find. Go through each day's listings (including times you are normally away from a TV set). Circle shows that seem uplifting or may have educational value. Find the time to read feature stories on current shows and personalities, blurbs on upcoming shows, brief resumes on the contents of the shows, and daily TV columns in the newspapers. Find specific talk show guests or specific themes that may benefit you. Watch for news features, special programs, educational classes, provocative talk shows and movies that deliver a timely message. When you have marked what is of value for the next week, you can plan nontelevision hours constructively. Many shows are repeated several times throughout one week or two-week periods. Find the most convenient times to watch what you choose.

Dare to enter the "unexplored land" of television. You will not regret it!

Passive TV watching is easier than reading. In both instances we sit and recline, but competent reading demands we keep the reading matter properly positioned, turn pages, and look up a word occasionally in the dictionary. For millions of people TV viewing has supplanted the vigors of reading. Passive TV watching is a shameful waste of a magnificent, magical medium. A good portion of the "prime time" offerings induce a subtle hypnosis, which if engaged in for enough years, can lead, ultimately, to a vegetation of the nervous system.

If after exploring all possible offerings and nothing is on TV that would benefit you, but you cannot bear to have the set off, go ahead and turn it on; but while it is on, try exercising, fixing the back door, planting seeds in pots, or some other indoor activity that would be of benefit to you. (Also examine why you are compelled to have the TV on constantly.)

If you walk or drive the same road home every day from school or work, you will never see what other avenues have to offer. Different paths offer new scenery, adventure and expanded horizons. Similarly, watching the same programs every evening can, without your knowing it, put you into a mental and/or psychic rut. What is preventing you from going beyond the pablum of regular "entertainment?"

A quick switch of the dial or click of the automatic changer can whisk you from the inane game show or situation comedy to inner and outer journeys, allegorical myths and legends, replays of spectacular Biblical events, dynamic historical narratives, or into the presence of the world's most famous and infamous personalities. Flick that switch and go deep down into mysterious ocean depths; directly onto the surface of another planet; or onto a live battleground.

ACTIVE WATCHING

Exercising while watching television will keep you fit and alert. Do exercises slowly, but with perfect form. Begin your exercises with roll-ups, which are similar to sit-ups. Keep the legs flat; reach the hands toward the feet; tuck the chin into the neck; raise the torso carefully from supine position, vertebra by vertebra, as if you were rolling up a carpet. Once you have "rolled" your limit, roll the spine back down, one vertebra at a time, to the supine position.

Leg dips are another great TV exercise. They are like squats, but have slow and perfect form. Balancing on the balls of your feet, squat, while keeping the spine erect. Push-ups are another effective TV exercise, especially during those wasted commercial minutes. Try standing on one leg while viewing TV, to improve balance. If you feel yourself fading during an important show, wrest your body free from that soft chair, and put it into "kamai," a slightly squat position, legs shoulder distance apart, viewing with a sharp, hawklike gaze.

TV viewing time provides a perfect opportunity to take care of the mountains of busywork that always seem to accumulate, such as ironing and mending. Spread newspapers on the floor to clean boots or auto parts, paint and renovate small objects or remove rust from golf clubs and outdoor tools.

Don't have time to go back to school? Would you like to travel but can't afford it? A magic box, a genie's lamp, the TV sits in your own home. Don't waste it!

Changing your attitude when watching TV from "go to sleep" to "go to school" will have enormous beneficial effects upon your mind.

TV can be more educational than going to college. If you cannot make it back to college, try converting your living room into a TV classroom. TV offers a diversity of information, relating to every aspect of our lives. Learn second languages, cooking, sewing, history, business, gardening, current environmental dangers, legal aid, household maintenance and more.

Professionals, students, serious artists and craftsmen, schoolteachers, dedicated homemakers and sportsmen are learning how to use the television as a working tool. TV is the single most phenomenal communications tool in recorded history. Any tool, if it is to be used constructively, must be understood and used properly.

SUGGESTED READING

Adler, Richard; Cater, Douglass, eds. *Television as a Cultural Force.* New York: Praeger Publishers, Inc., 1976.

Feshback, Seymour; Singer, Robert. *Television and Aggression: An Experimental Field Study.* San Francisco: Josey-Bass, Inc., Publishers, 1971.

Goodhardt, G. J. *Television Audience: Patterns of Viewing.* Lexington, MA: Lexington Books, Inc., 1975.

Logan, Ben T. ed. *Television Awareness Training: For New Awareness, New Decisions, New Action.* Woodstock, CN: Media Books, 1977.

Sherrington, Richard. *Television and Language Skills.* New York: Oxford University Press, 1973.

UNESCO, International Propaganda and Communications Service. *Television: A World Survey.* New York: Arno Press, Inc., 1972.

BICYCLES

The comeback of the bicycle is a sign of national sanity. In a time when our cities are saturated with automobiles, when noise and air pollution are increasing and the U.S. is running out of petroleum, an upsurge of interest in this best of the self-propelled vehicles is good news for everyone. For over a generation Americans have marched to the suffocating siren song of the combustion engine. We have spent billions to build dangerous, over crowded highways. Now the energy crisis is trying to tell us that we have pushed our gas-guzzling, stinking combustion machines as far as the limits of our resources will take them. The time has come to turn back to the bicycle, to shank's mare, to clean and conventional modes of public transportion.

Stewart L. Udall
former Secretary of the Interior

Currently, bicyclists across the country number about 75–100 million, making bicycling the number two leisure sport in the United States, surpassed only by swimming. The fantastic increase in the use of bicycles can be attributed to several factors. Some of these are the desire to travel ecologically and economically, to help keep the body healthy, and for recreational reasons. Both sales and accident statistics point out that big increases in bicycle use are within the 18-to-35 age group. The day of the bicycle being passed off as a child's toy is over.

Bicycles provide a self-sufficient mode of transportation, independent of fuel supplies, and an important survival tool. Bicycle-like devices may even be a future energy source.

Members of "civilized" societies need to take a long hard look at the way they transport their bodies. Don't our accepted modes of transportation have a relationship to our health?

Eugene Sloane in his *Complete Book of Bicycling* points out that if you get some sort of regular exercise, such as bicycling, you can expect to reduce chances of developing degenerative vascular diseases associated with heart attacks, strokes and high blood pressure.

You can also expect to be stronger and more resistant to injury, to sleep better, to be more relaxed in general, to live for up to five years longer, and, most importantly, to think better due to increased blood to the brain. Thus biking has the potential to relieve and reduce the anxiety and pressure associated with urban living.

MAINTENANCE AND REPAIR

To become a self-sufficient bike traveller, familiarize yourselves with bicycle maintenance and makeshift repairs. Learning repair and maintenance takes time, but gradually you will be able to repair all your bike's problems.

A properly-maintained bicycle can literally last a lifetime. It pays to know how to repair and maintain your bike correctly. Everything on your bike should be kept in proper adjustment. A well-maintained bike will last longer, be safer and more reliable.

Properly adjusted brakes are crucial to the safety of you and your bicycle. In addition to human injury, a head-on accident will warp the rim and bend the fork toward the frame. Because the fork is not designed for bending, it may weaken and crack when you bend it back to its original shape. Whenever you hit a bump, the fork's weak spot will be stressed. Continued stress can cause cracks, and necessitate replacement of the fork. See *Richard's Bicycle Book* for details on brake adjustment.

Check your tire pressure regularly. Lower tire pressure makes a bike harder to ride. Read the recommended tire pressure for your tire on the tire's sidewall. Bike racers use high-pressure "sew-up" tires, because they have less tire surface on the road and less rolling resistance.

The tire valve stem should always be perpendicular to the rim. A valve stem which is misaligned for a period of time will be cut by the valve hole in the rim, rendering your tube worthless. There is no effective way to repair a tube once the stem is broken or cracked. To prevent a broken stem, be certain the tires are properly inflated and the tube is inserted properly. A good pump is essential to biking self-sufficiency. Two superior brands are Zephal and Silca.

The chain should be cleaned and oiled regularly. Friction gradually wears down any piece of machinery, and lubrication is the best way to minimize friction. Clean the chain by soaking it in penetrant SS-1 (available at hardware stores) and scrubbing it with a toothbrush. Oil the chain only after you have cleaned it. Abrasive dirt will accelerate wear on the chain, chain wheel and rear sprocket. The best way to oil your chain is to place an individual drop of an oil-based lubricant on each roller, and then wipe. Repeat as often as needed.

Be sure your derailer is positioned directly above the sprocket you are engaged in or there will be undue strain on the gears. Lube the pivot points on the derailer often. Watch for frays or kinks on your cable and housing (they reduce efficiency), and replace if necessary.

All bearings (pedals, crank, headset and hub) need to be checked periodically, adjusted and lubed properly. If the bearings run dry, the bearing assembly will likely be ruined. The quick-release hub makes wheel and tire work easier, but it also makes tire theft a simple matter. You must decide which is more important.

A bicycle tool kit is essential. Some shops sell combination tools which incorporate three tools in one, such as a pliers, screw driver and crescent wrench. Combination tools help reduce your carrying weight. A compact tool kit (purchased at any bike store) usually contains a screw driver, crescent wrench, spoke wrench for truing wheels, tube kit with repair kit and tire irons, and a container of lubricant. To get an idea of how your bicycle works, observe it. Set it up on a stand upside down and watch all the moving parts.

PEDAL POWER

A bicycle can take you three or four times faster than walking but uses five times less energy. Properly maintained, a bicycle should well outlive its owner. Requiring no gasoline, it is an excellent alternate mode of transportation, especially when fuel becomes scarce and/or exorbitant in cost.

An unloaded bike travelling briskly along at about 20 miles per hour could make it to San Diego from Los Angeles or New York to Philadelphia in about six hours, although you would have to be in top shape to attain that speed. Before the automobile was around, traversing such distances would have taken much longer.

The New Jersey-based Pedalpower Company specializes in battery-operated electric-drive systems for bicycles which assist the rider in going up hills, extend cruising time and give the rider rest periods. A pedal power unit can be attached to any bicycle, and the company makes their own bikes. This bike uses no gas and requires no license to operate.

The Pedalpower tricycle is an effective work tool for hauling small loads too difficult to carry on a bicycle or motorcycle. It is also very useful for the handicapped or elderly. Distances of up to 25 miles can be travelled per battery charge, with a speed of about 8 miles per hour without pedalling, and up to 15 miles per hour pedalling.

Bicycles are effective tools. They can carry large loads, up to 200 pounds and more (if properly balanced). A wide and versatile selection of packs are available for biking. A bike trailer can carry about 80 pounds or two small children. Thirty-one-year-old Wiegand Horst Lichtenfels travelled 47,500 miles around the world on a fully-packed bicycle weighing 300 pounds. The three-year trip led Lichtenfels through 42 countries.

In the late 1880s there was a revolution in both the home and industry in the United States as pedal-powered devices became widely adapted to a wide array of jobs formerly done by hand. However, as the internal combustion engine gained in use and popularity, pedal power was slowly forgotten.

Possibly one of the most revolutionary devices of the century has been developed by Rodale Resources in Pennsylvania. Rodale Resources is a division of Rodale Press, the people that publish *Organic Gardening, Prevention* and *Bicycling*. The device is called the *Energy-Cycle Workhorse*, a pedal-powered unit that enables an average person to generate up to 0.3 horsepower for long periods of time, while pedaling at 70 to 90 rpm. The concept of pedal power is not a new one. At the turn of the century many machines were designed to be operated by pedal power or some sort of human energy (the treadle sewing machine).

With this pedal-powered unit, homesteaders and urban survivalists sit in an adjustable padded seat. Pedaling at an easy pace, they can generate more than enough mechanical energy to operate kitchen appliances and power tools efficiently.

The cycle is designed to give the operator excellent leverage on the pedals. Power is transmitted to the output shaft via a sprocket-and-chain mechanism. Between jobs, the operator can "shift gears" by moving the chain from one pair of sprockets to another.

In "low gear" the energy cycle develops the high torque and low rpm required for heavy jobs, such as shredding compost; in "high gear" it develops lower torque but higher rpm, for jobs like polishing metals, where speed is more important than brute force. With the "gear shift" the operator can always pedal at 70 to 90 rpm, an average pedaling speed for most people.

The high-muscular, energy-conversion efficiency makes pedalers significantly more productive than with other muscle-powered systems. With the appropriate attachments, kitchen chores can be done quickly and efficiently, including milling flour from whole grain; grinding, chopping and blending foods; even churning butter and making ice cream. Milling enough flour for six loaves of whole wheat bread takes about 20 minutes. The same task may take much longer when a mill is powered with a hand crank. The cycle leaves both hands free.

Moved into the workshop, the cycle can power small lathes, drill presses, saws and other tools to cut, shape and finish metals, woods and plastics. It will energize potter's wheels, jeweler's lathes, polishing wheels, lapidary equipment and sculptor's tools.

Outdoors the cycle can help pump water for irrigation, saw wood, split logs, and when equipped with a winch, pull stumps. Rodale Resources also produces the Energy Cycle Mechanical Mule model, a portable pedal-powered system designed to help home gardeners plow, harrow and cultivate in a fraction of the time required to do the same jobs with hand plows, spades, rakes and hoes.

After the day's chores are done, the cycle, equipped with an electrical generator can furnish enough power for radio and television sets. It can charge storage batteries and even furnish power for lighting during emergencies and blackouts.

Pedalling the cycle provides good exercise and is ecological. Cycling burns up excess calories, trims unsightly flab and develops strength and stamina. The potential health benefits of regular sessions on the EnergyCycle have interested physicians and physical therapists. Patients may recover faster from illness or surgery if they get regular exercise. Patients can follow the doctor's "prescription" to get exercise without having to leave their urban wilderness home. Provided with the incentive to power a favorite hobby or energize the television set, these patients have an incentive to exercise.

Thus the EnergyCycle not only provides an excellent physical workout, but it also provides your home with an ecological source of "human-powered" electricity.

Maybe in the years to follow we will see a resurgence in the use of pedal power. The technology has been here for a long time, waiting only to be applied. We as a nation can begin the transition now, or wait for an emergency, such as a major gas crisis, and be forced to accept alternative energy sources. Which is the wiser choice?

SUGGESTED READING

Ballantine, Richard. *Richard's Bicycle Book.* New York: Ballantine Books, 1972.

Belt, Forest; Mahoney, Richard. *Bicycle Maintenance and Repair: Frames, Tires, Wheels.* Indianapolis: Theodore Audel and Company, 1974.

Coles, Clarence W.; Glenn, Harold T. *Glen's Complete Bicycle Repair Manual: Selection, Maintenance, Repair.* New York: Crown Publications, Inc., 1973. A photo-illustrated, complete and extensive manual.

Kolin, Michael J.; de la Rosa, Denise M. *Complete Cycling: All About Frames, Frame Builders and Bicycle Setup.* Emmaus, PA: Rodale Press, 1979. What to look for when purchasing a bike, how to choose the right frame, design and components. Tips from twenty top bike builders.

McCullagh, James C., ed. *Pedal Power.* Emmaus, PA: Rodale Press, 1977. An impressive coffee table volume that also includes the specifics on generating pedal devices. Also includes the history of pedal-powered devices. A very good book.

Sloane, Eugene. *Complete Book of Bicycling.* New York: Trident Press, 1970.

V
CITY SURVIVAL

Is it realistic to expect that in the event of wide-scale natural or human-made catastrophes we can "head for the hills" as was suggested in the movie "Earthquake?" Who among us would survive if we ran into the hills? Who among us in the large sprawling modern cities has the option of heading for the hills? Escape is generally not an option, except for a small minority. And the question is: escape to what?

Wilderness survival and city survival are one and the same. The techniques and knowledge displayed in survival manuals can be applied in the city in the event of large-scale disaster. It's that simple.

Wilderness survival instructors have begun teaching people how to survive in the wilderness in the event of an urban crisis, breakdown or massive civil disorder. They want to train aware individuals to survive in the wilderness, so that if urban crises occur, they would be prepared to adapt and survive. We can spread the word, and teach others basic survival skills. We must be able to rely on ourselves, become more self-sufficient and less dependent on faulty city energy supplies.

Those who choose to stay in the urban wilderness must accept the negative influences with the benefits. Ostrich-like we ignore negative realities. Passive attitudes avoid crises, but result in unpreparedness when disasters strike. We must view ourselves as astronauts in space, captains on ships, or travellers in the desert. In each instance a specific set of safety or survival practices and attitudes must be adopted and become instinctive. We dwellers of the cities must adopt a new posture of awareness and willingness to take the required safety precautions—to survive into the future.

"Be prepared," says the Boy Scout motto. Being prepared means having considered the possibility of problems arising, and having decided to take specific preparatory action, if and when crises occur. Taking a first aid course and learning how to repair your bike are good examples of being prepared. Do not expect disaster; but be ready for it.

FIRST-AID

Basic first aid knowledge is a must for anyone who intends to be prepared for city and wilderness emergencies. The American National Red Cross offers free first aid classes in virtually every population center in the United States. Take advantage of these valuable free lessons in your community.

Most fatalities from accidents can be prevented by acting quickly and accurately. Better yet, eliminate accident-producing situations; be aware of danger and stay out of the way. I recommend having several up-to-date first aid charts and manuals. *Standard First Aid and Personal Safety* by the Red Cross is excellent. Place manuals strategically around the house, the yard, the office, the car and in the backpack. Refer to these books often, so that you will be ready to act when emergencies occur. Beware of thoughts such as, "I don't want to think about injuries and blood." This type of thinking renders you totally useless in dealing with your own or loved ones' emergency situations. Don't expect accidents; but do anticipate their possible sudden arrival with thoughtful preparation.

SHOCK

Many types of injury or disease may result in shock or shocklike states. Shock is the major hazard in any accident, and all accident victims should be treated for shock *immediately*.

Evidence of injury may or may not be readily apparent. Usually, the accident victim is conscious, but may be confused. The most frequent complaint is thirst. Restlessness is also common. Low blood pressure, weak pulse, cold clammy extremities, increased respiration and a pale face beaded with cold perspiration are symptoms of advanced shock. Persons in shock are in danger of sudden death and must be watched closely.

Insist the victim lie down. Cover the body (even on warm days) to prevent the loss of body heat. Do not apply external heat unless there has been exposure to extreme cold. It is best if the extremities remain cool.

BURNS

Burn injuries from sun, wind, fire, electricity or chemicals pose a constant threat, both in and out of the home. The chance of being burned is the same whether you're backpacking, working or staying around the house. Most of us are burned, in one way or another, with regular frequency; the reason we forget most of our burns so quickly is that the "degree" of severity is insufficient to bother us for very long. The degree of burns is classified as first-degree, second-degree or third-degree.

First-degree burns result from overexposure to the sun, or brief contact with hot liquids and hot objects (120° or over). Redness or discoloration and mild swelling accompanied by pain, but rapid healing, is indicative of a first-degree burn. Damage is limited to the outer layer of the epidermis.

Second-degree burns result from considerable overexposure to the sun, or extended contact with hot objects or liquids (140° or over). These burns form blisters, considerable swelling that lasts for a period of days, with persistent pain (since nerve endings in the skin have been destroyed) and slow healing. Damage extends through the epidermis and to the dermis skin layer.

Third-degree burns are caused by prolonged contact with liquids or objects (160° or over), resulting in tissue destruction and white or charred skin. These burns often resemble second-degree burns at first, but both the epidermis and dermis of the skin have been destroyed. New tissues may form but often skin grafting is required.

BURN TREATMENT

FIRST-DEGREE BURNS

1. Instantly immerse the burned area in cold water or cover with ice or snow.
2. After five minutes, inspect burned area(s) carefully and return to water for five minutes, if still brightly inflamed.
3. When redness has subsided, cover with a clean dry dressing for no more than five minutes.

SECOND-DEGREE BURNS

1. Immerse burned area in cold water or cover with ice or snow.
2. After five minutes, inspect the burn and return to water for another five minutes, if necessary.

3. Cover with clean cloths that have been wrung out in ice water; wring cloths in ice water as soon as they become warm, and repeat.
4. Continue step 3 until bright redness subsides.
5. Gently blot the area dry and apply dry sterile gauze or other clean cloth for protection. *Do not break blisters or remove any skin tissue.*
6. Keep burned arms or legs elevated.

THIRD-DEGREE BURNS

1. *Do not remove any particles of charred clothing that may have adhered to the skin.*
2. Cover the burns with thick sterile dressing.
3. Elevate burned hands and feet.
4. *Do not immerse burned areas in cold water or apply snow or ice;* cold may intensify the shock reaction.
5. Apply a cold pack to the forehead, back of neck and hands or feet, if these areas are not burned.
6. *Rush the burn victim to the hospital as soon as possible.*
7. If circumstances prevent hospitalization for an hour or more, have the victim sip the following solution: 1 teaspoon salt and ½ teaspoon baking soda, dissolved in 1 quart water. An adult should sip steadily at the rate of 4 ounces for every 15 minute period. *Do not administer to a victim who is semiconscious and unable to swallow properly.*

CHEMICAL BURNS

1. Remove all clothing from the affected area.
2. Wash away the chemicals with running water; use a hose or shower if available; continue for at least five minutes.
3. If possible locate the chemical that caused the burn, and follow whatever first aid directions are given on the container.
4. Apply a clean, dry dressing.
5. *Get medical help immediately.*

CUTS AND ABRASIONS

Keep in mind that a person can bleed to death in one minute or less: *STOP THE BLEEDING!!!!* Even if you have to apply direct pressure with your unwashed hands, get that bleeding stopped! If direct pressure fails to arrest the bleeding, you must apply pressure *instantly* to the artery that supplies blood to the injured limb. Protecting the wound from infection is of secondary importance, only after the bleeding has stopped.

For minor cuts, punctures and abrasions where bleeding is not severe; wash the wound thoroughly with soap and water. Apply a clean, dry dressing or bandage.

For major wounds direct pressure is the only safe, sure emergency aid. If you use cloth to stop the bleeding, bind it into place.

Pressure points on the arms are on the inside, in the groove or hollow between the biceps or triceps. *You must press hard against the skin until you feel you're pushing against the arm bone.* Readjust slightly until all bleeding stops. Pressure points on the legs are on the inside of the thighs, high up in crotch area. *Press hard upwards against the pelvic bone, until all bleeding stops.* As soon as the heavy blood flow has been stopped, or reduced significantly, insert a clean cloth between your hand and the wound to keep it free from infection. Elevate the body where the wound is located.

Tourniquets should be used only if all else fails to halt the bleeding. Use a tourniquet only if you are willing to sacrifice the limb in order to save the life. Tighten a tourniquet very slowly—just enough to slow the bleeding to a trickle.

Have only a doctor or other competent medical person loosen the tourniquet. If a tourniqueted victim must be covered while awaiting medical attention, be sure you tell the medical attendant about the situation the moment they arrive. If you must leave a victim in a tourniqueted condition, place a note prominently on top of the covering telling the location of the tourniquet and when it was applied. *The victim must be treated for shock before you leave to summon medical assistance.* This is a crucial point that is often overlooked, even by "experts."

SUGGESTED READING

American National Red Cross. *Advanced First-Aid and Emergency Care.* New York: Doubleday, 1973.

_____. *Standard First-Aid and Personal Safety.* New York: Doubleday, 1973. Both Red Cross texts come in paper for portability; both are excellent and often used in first-aid courses.

Boy Scouts of America. *First-Aid Skill Book.* (Teacher's Guide) North Brunswick, NJ: 1974.

Brown, Robert E. *Hip-Pocket Survival Handbook.* Bellevue, WA: American Outdoor Safety League, 1979.

Darvill, Fred. *Mountaineering Medicine.* Seattle: Mountaineers, 1978.

Mitchell, Dick. *Mountaineering First-Aid: A Guide to Accident Response and First-Aid Care.* Seattle: Mountaineers, 1978.

Rothenberg, Robert. *First-Aid: What to Do in an Emergency.* New York: Crown Publishers, 1976.

Wilkerson, James A. *Medicine for Mountaineering.* Seattle: Mountaineers, 1975. Mountaineering medicine at its best.

EMERGENCY WEATHER

Changing weather patterns place the urban dweller in a precarious position. Severe winters in the Northeast and Midwest, an increase in flooding, drought conditions in the Southwest, and an increase in hurricane, earthquake and tornado activity make it a necessity for every city dweller to know specifically how to cope with emergency weather conditions. Being prepared will make your life easier in the long run.

BLIZZARD

A blizzard is the most dangerous of all winter storms. It combines cold air, heavy snow and strong winds. Moving, blowing snow often reduces visibility to only a few yards.

Be prepared for isolation. Make sure you could survive at home for a week or two, if a storm isolated you and made it impossible for you to leave.

Keep an adequate supply of heating fuel on hand and use it sparingly; your regular supplies may be heavily curtailed by storm conditions. If necessary, conserve fuel by keeping the house cooler than usual, or by closing off a few rooms temporarily. Keep some kind of emergency heating equipment and fuel, so you could keep at least one room of your house warm. A camp stove and fuel, or a supply of wood is fine if you have a fireplace or wood stove. If your furnace is controlled by a thermostat and your electricity is cut off by a storm, the furnace probably would not operate, and you would need emergency heat.

Also keep an emergency supply of food and water, as well as emergency cooking equipment. Keep some food around which does not require refrigeration or cooking, such as dried fruit and nuts, whole wheat bread, canned tuna and peanut butter.

Make sure you have a battery-powered radio and extra batteries on hand, so that if your electric power is cut off, you could still hear weather forecasts and information broadcasts by local authorities. Flashlights or lanterns are also essential.

EARTHQUAKE

If an earthquake occurs in your area, keep calm; don't run or panic. If you take the proper precautions, you shouldn't get hurt.

Remain where you are. If you are outdoors, stay outdoors; if you are indoors, stay indoors. Most injuries occur from falling walls or electrical wires, as people are entering or leaving buildings.

If you are indoors, sit or stand against a wall (preferably in the basement) or in an inside doorway. Take cover under a desk, table or bench (in case the wall or ceiling should fall). Stay away from windows and outside doors.

If you are outdoors, stay away from overhead electric wires, because they could shake loose and fall. If you are driving an automobile, pull off the road cautiously and stop as soon as possible. Remain in the car until the disturbances subside. When you drive on, watch for fallen objects, electrical wires and broken or undermined roadways.

TORNADO

When a tornado warning is issued, take shelter immediately. The warning means a tornado has actually been sighted, and may strike in your vicinity. You must take action to protect yourself from being blown away, struck by falling objects, or injured by flying debris. Your best protection is an underground shelter or cave, or a substantial steel-framed or reinforced concrete building.

If you are at home, go to your underground storm cellar or basement fallout shelter, if you have either. If not, go to a corner of your home basement and take cover under a sturdy workbench or table, but not underneath heavy appliances on the floor above.

If your home has no basement, take cover under heavy furniture on the ground floor in the center part of the house, or in a small room on the ground floor, away from outside walls and windows. As a last resort, use a nearby ditch, excavation, culvert or ravine for protection.

Doors and windows on the sides not facing the tornado may be left open to help reduce damage to the building, but stay away from them, to avoid flying debris. Do not remain in a trailer or mobile home, if a tornado is approaching; take cover elsewhere.

If you are in an office building, go to the basement or to an inner hallway on a lower floor. In a factory, go to a shelter area or the basement.

If you are outside in open country, drive away from the tornado's path, at a right angle to it. If there isn't time to do this, or if you are walking, take cover and lie flat.

HURRICANE

If your house is on high ground, and you haven't been instructed to evacuate, stay indoors. Don't try to travel; you will be in danger from flying debris, flooded roads and fallen wires. Tape an "X" across windows with masking tape or other strong tape to prevent breakage. Secure any loose objects around your home, or take them inside.

Listen to your radio or television set for information and advice. If the center or "eye" of the hurricane passes directly over you, there will be a temporary lull in the wind, lasting from a few minutes to perhaps a half hour or more. *Stay in a safe place during the lull.* The wind will return—perhaps with even greater force—from the opposite direction.

FLOOD

In many areas, unusually heavy rains may cause sudden flash floods. Small creeks, gullies, dry streambeds, ravines, culverts and grounds can flood quickly and endanger people, before any warning can be given.

Be aware of this hazard during heavy rains and be prepared to protect yourself. If you see the possibility of a flash flood, move immediately to a safe location. Don't wait for instructions to move. Notify local authorities of the danger, so that other people can be warned.

COLD WATER SURVIVAL TEST

1. What is the best, most dependable survival tool?
2. What naturally insulates the body?
3. Immersion hypothermia results when the body core temperature goes below 98.6°. True or False?
4. Does water or air drain heat from the body faster, and at what rate?
5. At least a 2/3 increase in survival time could be predicted for persons wearing proper clothing made of what fabric?
6. If you are in cold water, keep warm by moving rapidly. True or False?
7. Hypothermia can result without total immersion in water. Name two ways.
8. Is there any ideal posture to adopt when in a cold water survival situation?
9. You are in very cold water and the boat turns over. What is the best thing to do? A. Stay with the boat; B. Move about rapidly to stay warm; C. Swim for shore, about two miles away.

10. What technique can be used to warm the body without obvious physical movement?

Answers:
1. The brain.
2. Fat.
3. False. Immersion hypothermia results when body temperature goes below 78°.
4. Water drains heat from the body, 20 to 240 times faster than air.
5. Wool. Virgin wool has the ability to insulate even when wet.
6. False. Moving about rapidly increases blood circulation. This causes the blood to carry heat from the body's core to the extremities and out of the body. A person with bad circulation would have a better chance of survival in this instance.
7. Rain, snow, wind, wet feet and wet hair are all possibilities.
8. Yes. The heat escape lessening position or foetal position. You can stay in this posture only if you are wearing a floatation jacket.
9. A. Stay with the boat.
10. Visualization. Visualize drinking scalding hot tea. Pretend to sip and feel the hot warming tea as it goes down the esophagus and into the stomach. Feel the heat spreading from the center of your body to the entire torso.

SUGGESTED READING

Accerano, Anthony J. *Outdoors Emergency Manual.* New York: Winchester Press, 1977.

Angier, Bradford. *How to Stay Alive in the Woods.* New York: Macmillan Publishing Company, Inc., 1962. A very good nontechnical guide to survival.

———. *Survival with Style.* New York: Random House, Inc., 1974.

Barker, Harriett. *The One Burner Gourmet.* Maple City, MI: Great Lakes Living Press, 1975. A very versatile cookbook for camping stove cookery. Just because you don't have a kitchen doesn't mean you have to sacrifice a good meal.

Bridge, Raymond. *The Complete Snow Camper's Guide.* New York: Charles Scribner's Sons, 1973. A handbook for winter and wilderness travel. Helpful for learning how to get along under snow conditions without the conveniences we are accustomed to.

Danielsen, John. *Winter Hiking and Camping.* Adirondack Mountain Club, 1977. A basics primer for winter backpacking skills, general camping know-hows and keeping calm under prevailing cold conditions.

Eng, Evelyn; Garb, Solomon. *Disaster Handbook*. New York: Springer Publishing Company, Inc., 1978.

Fear, Eugene H. *Surviving the Unexpected Wilderness Emergency: A Text for Body Management Under Stress*. Tacoma, WA: Survival Education Association, 1974.

Fletcher, Colin. *The New Complete Walker*. New York: Alfred A. Knopf, Inc., 1974. A thorough discussion of equipment: how to pick it, how to use it. This well known backpacking book has become a classic in the field.

Hart, John. *Walking Softly in the Wilderness: The Sierra Club Guide to Backpacking*. San Francisco: Sierra Club Books, 1977. A handbook that emphasizes ecological use of the outdoors.

Keatine, W. R. *Survival in Cold Water: The Physiology and Treatment of Immersion Hypothermia and of Drowning*. Philadelphia: J. B. Lippincott, 1969.

Kelsey, Robert J. *Walking in the Wild: Complete Guide to Hiking and Backpacking*. New York: Funk and Wagnalls Company, 1973.

Manning, Harvey. *Backpacking One Step at a Time*. New York: Random House, Inc., 1973. A comprehensive and ecological view of backpacking. Very readable.

Olsen, Larry D. *Outdoor Survival Skills*. New York: Simon & Schuster, 1975. A very useful book by an important wilderness expert. Provides a storehouse of information, aimed primarily at the Western United States audience, for their particular weather conditions.

Szczelkun, Stefan A. *Survival Scrapbook 2: Food*. New York: Schocken Books, Inc., 1973.

SHELTER

All too often we take our comfortable homes for granted. They can be wiped out from an earthquake, fire, flash flood, tornado or other natural disaster. In the aftermath of an urban disaster you may be able to use the remainder of your home as a temporary shelter, providing it is not under water or burned to the ground.

If you cannot inhabit your house or apartment, three possibilities are open to you: find a shelter ready to occupy for the night or longer; make a shelter from a tent or tarps; or build a shelter from natural ingredients.

Although building one's own wilderness shelter has so often been romanticized, in a survival situation it is far more practical and realistic to find a shelter ready to occupy. In some instances, finding an emergency shelter can be as simple as going to a neighbor's house or youth hostel, local church, school or bomb shelter. Sheltered picnic areas in parks and the hollow tubes often found in playgrounds both make excellent temporary shelters. A freeway bridge will protect you from rain, although there is nothing glamorous about it.

MAKESHIFT SHELTER

Small abandoned shelters can be found in most areas; old wooden shacks and shanties, and even some stone or sheet metal shacks. Abandoned railroad cars have become permanent homes in Mexico, so obviously they will work as temporary shelters.

The automobile provides acceptable shelter, providing it does not leak and can easily be cleaned of any broken glass. The doors, hood and seats can be used with other materials to create a much larger and more comfortable shelter. Car mirrors can be used as signalling devices, and for starting fires. Motor oil will work as a type of candle fuel. Seat padding makes excellent tinder for a fire. Hub caps come in handy as cooking pots for boiling water and cooking food; they will also work as shovels to dig holes for a toilet, post holes and storage holes.

If you are far from civilization, consider finding a cave. Caves are not difficult to find, and make good living quarters for a day, a week or longer. First inspect for deep pits and leaking water. You will need to consider the possibility of collapse with human-constructed caves. Natural caves in solid rock are usually structurally sound. A large hollow tree stump, big enough to fit your body lying down, would be acceptable. Inspect the tree for ants, bees, an abundance of insects, and snakes, before moving in. The base of an old fallen tree will make a suitable shelter. A toppled tree often leaves a nearly vertical wall of dirt-encrusted roots, which serves fine as one wall of your shelter. Lean poles against this wall at a 45° angle, and then cover with a tarp or other suitable material.

If you have a tent, tube tent, tarp or even a grommetted ground cloth, you can build a temporary shelter in your backyard. They can be fashioned into a variety of shelters, depending on your needs, skills, and how quickly you need to get under cover. Practice now, with preparedness—not disaster—in mind.

The tube tent is an excellent temporary shelter that you would do well to purchase and have on hand. It is simply a large tube of thin plastic, about seven feet long, and costs about two or three dollars. Put a rope through the tube, and tie it at both ends to two trees, or anything sturdy. As you sleep on the bottom, the tube tent forms a triangular shape. The tube tent will sleep two comfortably on the ground; it folds down easily and takes up little space.

One of the most versatile items is a lightweight, waterproofed nylon tarp. Fashion a tarp into a shelter, or use it as a ground cloth. An ideal size is 10' x 12' with grommet holes and ties. Made of rip-stop nylon, a tarp this size will weigh 2½ to 3 pounds, and will cost in the neighborhood of $30. Compared to a tent, a tarp is lighter, far less expensive, less bulky and more versatile.

If there is no danger of rain, but you want to protect yourself from the night dew, use the tarp in one of several ways. Lay it down like a ground cover with your sleeping bag on top. If the tarp is big enough, pull it over the top section of the sleeping bag. Pine needles, leaves, newspapers or rags will give you a cushion over which to place a blanket. If it is raining, or rain appears likely, dig a trench around your tent with an outlet, so the water will run away from the tent. The use of the tarp is only as limited as your own imagination and ingenuity.

Before building a temporary home, select the most desirable site which offers protection from the prevailing night winds, storms, and is out of the way of potential flash floods, falling rocks or high tide. The location should be level enough for your bed and your fire pit, and should be on dry ground, away from green grassy areas and creek bottoms. Camping near trees is alright, but do not camp directly under large trees. Rain and dew compound the dangers of lightning and falling branches.

PERMANENT SHELTER

An abundance of building supplies is required for a permanent shelter. Water and firewood should be readily available. Make the shelter strong enough to hold up to high winds or heavy snowfall. Use strong main poles, and secure them firmly. To lash them together (if you have no twine or fishing line), use cattail or yucca leaves, grape vines, or whatever material is both available and strong. Make the shelter large enough to contain a fire and provide ample space for movement around the fire. You will need to consider space for your equipment and an inside wood pile to last the night.

Constructing and erecting a tipi is not that difficult a job, and a tipi is practical in any climate. Lash three long poles together and prop them up as a tripod. The length of the poles will determine the tipi's floor space; the longer the poles the larger the space. Place more poles through the lashing to frame out the tipi, and cover it (preferably with skins or tarps that have been sewn together specifically for a tipi cover). Traditionally, tipi entrances face east to catch the rising sun.

The fire pit should be just inside the entrance, but make sure it will not interfere with entering and exiting. Dig a small hole for the fire, line it with stones, and prop up a large stone behind the fire as a heat reflector, for maximum warmth. Use a layer of grass or other available leaves for your floor matting, but keep clear of the fire.

Another more elaborate shelter design is the lean-to, which is supported by two trees or two vertical stakes. Lash a pole between two trees at the maximum height of the shelter. Attach two more poles to each tree, perpendicular to the crossbar, and at approximately a 30°–45° angle to the ground. These will form the sides of your shelter. Now you've got your frame finished. Secure more cross beams to frame out the sloping roof and the triangular sides. Use leaves, branches, mud and whatever is around to complete the shelter. Sheets of bark or large leaves, covering the shelter in layers from the bottom up, will make the shelter virtually waterproof.

The dome is another possible shelter. Begin by choosing a site, and draw a circle on the ground for the proposed circumference. Drive vertical poles of sufficient length along this circle about one foot apart or closer. Using pliable young branches, curve them inward and tie them together at the center top. After attaching all the poles at the top, you can attach crossbars for a smoke vent and begin to cover the dome with available materials. In snow country, this shelter can be the frame for a type of igloo.

Many shelters are possible. Your only limits are your own ingenuity, energy, and available building materials. When finished, leave the lean-to standing. Made of natural material, it will not look out of place and may prove helpful to someone else at a future time.

Learn how to build these temporary dwellings, or at least to put up a tent, with ease. Go camping, backpacking or crosscountry cycling. All self-sufficient skills are interrelated. Learning survival skills and employing them provides ample preparation for city emergencies and general adaptability in the urban wilderness.

SUGGESTED READING

Boyle, Godfrey; Harper, Peter, eds. *Radical Technology.* New York: Pantheon Books, 1976. Sections on food, energy, shelter, clothing, autonomy, communications, tools and communities. Good section on temporary shelters. Many illustrations and photographs.

Faegre, Torvald. *Tents: Architecture of the Nomads.* Garden City: Anchor Press/Doubleday, 1979. Every conceivable type of tent imaginable and how to construct it. Specializing in tents that harmonize with the environment.

Laubin, Reginald and Gladys. *The Indian Tipi.* Norman, OK: University of Oklahoma Press, 1977. A top-quality book with black and white photographs, color photographs and line drawings. Covers history, symbols and the how-tos of making your own tipis.

Szczelkun, Stefan A. *Survival Scrapbook 1: Shelter.* New York: Schocken Books, Inc., 1973.

FIRE

One night my friend Jim Graf told me about an experience he had with fire. "Too often," Jim said, "new hikers build indiscriminate fires in the damndest places! I've seen the remains of campfires that had been built in the worst locations. I once started a fire accidentally with a cigar butt. I was ashes and elbows putting the damn thing out with foot stompings and handfulls of dirt. It burned about a twenty foot radius before I stopped it. My hands were cut to ribbons. I've seen fires start from geologists' hand lenses, cigarettes, flares, lightning, fireworks and plain arson. We've really got to exhibit care and finesse when we encourage others to try these wilderness skills."

Fire, long evoking the imagination of humankind, is both a necessary tool and a feared enemy. Fire warms the night, cooks food, dries clothes, signals for help, purifies water, makes tools and lights the way. Controlled, fire is a welcome companion; uncontrolled, fire can wreak disaster and leave a trail of desolation. Fire, like knives, money, or knowledge is neither good nor bad in itself. It just *is*. Our use or misuse of fire determines our positive or negative connotations.

FIRE STARTERS

Every wilderness or city traveller should carry some means for starting fire. A butane lighter takes very little space, is lightweight and is a sure light. These lighters cost under $2 and last for quite a while. The butane will be used up eventually, but these lighters are worthwhile investments for short-term survival situations. Refillable lighters tend to dry out rather quickly, so unless you carry fuel along, they are acceptable for only short time periods.

Common wooden stick matches are the best matches to carry. Store them in a solid waterproof and unbreakable container. A rough surface on the outside or a small piece of sandpaper on the inside guarantees a good lighting surface. Wooden matches can be dipped in paraffin to protect them from moisture. Look for windproof and waterproof matches in camping stores everywhere. Camping equipment manufacturers have attempted to conquer the problem of moisture with innovative waterproof match containers, waterproof stick matches and matches with extra-long heads which insure a light even under the most difficult weather conditions.

A magnifying glass will start a fire, help locate splinters or study insects. As children most of us started fires with a magnifying glass. A camera lens can be used in the same manner. If your camera does not have a removable lens, take out the film and open up the back of the camera, to focus the sun's rays directly through the camera and onto the tinder. If you start a fire with a magnifying glass or a lens, use dry, fine tinder and blow lightly and steadily into the coals as soon as they appear. Gradually add more tinder and blow harder until the tinder bursts into flames.

The bulb reflector in a flashlight will work as a fire starter. The curved surface of the reflector focuses the sun's rays onto the tinder, quickly igniting it. Remove the bulb, insert tinder through the bottom opening, and aim at the sun.

The friction method of starting fires requires disciplined effort to achieve success. It involves spinning a drill with a bow into a foot piece, to create enough heat to start a fire.

MAGNESIUM FIRE STARTER

By far one of the most significant developments in fire starting devices has arrived in the last decade. The magnesium fire starter is a solid block of magnesium which measures 3" x 1" x 3/8" and weighs less than two ounces. The magnesium fire starter has a sparking device along one side of the block. There is a hole in the tool so that it can be carried easily on the keychain; attached to a keychain, the fire starter is light and unobtrusive. One is ready for survival situations, whether at a business meeting or in the mountains.

Scrape off a small pile of magnesium shavings with your pocket knife. Because the magnesium is soft, producing an adequate pile of shavings should take no more than 20 to 40 seconds. Bunch up the magnesium shavings into a compact pile and place into some tinder. Holding the fire starter tool in one hand, briskly scrape the flint-composition sparking insert to produce sparks. Hold the tool firmly while doing this, or you may scatter the pile of shavings. When sparking, hold the tool close to the magnesium shavings so that the sparks can be directed into the shavings. When you have mastered the art of sparking with this tool, you'll be able to ignite shavings in under 10 seconds.

I carried one of these tools on my keychain for about nine months. I estimated that I got close to 300 lights, and with only a slight indentation on the side of my tool. Eventually, the tool itself wears away from continually scraping off bits of magnesium. My intensive use of the tool for a nine-month period wore off no more than 5% of the magnesium. I suspect that one tool is worth thousands of lights and will literally last a lifetime of normal use.

To find out if the sparking insert on the magnesium fire starter is made of the same material as the metal match (a rodlike device), I phoned the manufacturer. I learned that the sparking insert on the magnesium fire starter is totally different in composition from the metal match. When I pressed for specifics, the manufacturer told me that the actual ingredients of the sparking insert is a proprietary secret, but that the sparking insert of the magnesium fire starter is definitely much harder than the metal match and should last a lifetime. In my own observations of one year, I have seen no powdering on the magnesium fire starter's sparking insert.

I have tested both the magnesium fire starter and the metal match in similar conditions. While the magnesium fire starter can produce a fire in virtually any tinder (even damp), the metal match must have ideal dry tinder (extremely dry grass, cotton or cattail down) to be effective. Several times I have been in situations where I needed to produce fire and had only a metal match. Once, I was lucky and found an old mattress someone had dumped over a mountain hillside. The cotton stuffing easily ignited with the metal match. In another incident with the metal match I was not so lucky. The main flammable ingredient in the area was dry grass. Although I finally did produce a flame with dry grass, I wore down 45% of the metal match in the process. With a magnesium fire starter, I've been able to produce fire quickly in most situations. Don't think you'll never need to know this skill simply because you live in the city.

You can still use the magnesium fire starter when travelling in damp and wet weather. A wet fire starter is no less effective than a dry one, because the shavings are highly flammable. Even damp (but not soaking wet) material can be ignited with the magnesium fire starter; it just takes a larger pile of magnesium shavings. The manufacturer claims that the magnesium ignites with a 5400° flame. These points make this tool valuable to boaters, rafters, winter and wet weather hikers and campers.

A concerned lady once asked me if a magnesium fire starter could explode (since magnesium is highly flammable). I did not know the answer. To find out, I tossed my magnesium fire starter into a camp fire one evening. In about 10 to 15 minutes, I saw a deep golden glow, a small sparkler effect, and it was over in five minutes: a beautiful fire but no explosion.

This tool is designed to be carried on your key chain. The main flaw that I have discovered is with the chain coming open. You will need to carry a pocket knife to use on the fire starter. I suggest a small inexpensive knife under $2 (such as the Christy knife) that you use solely on the fire starter and other small jobs. Lacking a knife, it is possible to use the firestarter with some other piece of metal, a sharp stone or shell; it is just more difficult that way.

I regularly conduct urban nature hikes where I teach plant uses and survival skills, and so I have regular opportunities to show and demonstrate this unique tool. Occasionally, someone will say, "This tool doesn't work!" The tool is a piece of magnesium with a sparking insert—it always works. Failure to produce a flame is usually a result of either not shaving off enough magnesium shavings, not pushing all the shavings into a single pile before attempting to ignite, or not holding the tool firmly when igniting, causing your hand to slip and scatter the shavings.

The magnesium fire starter is a simple, compact, lightweight, effective, easy-to-use, all season, fire starting tool. I feel it won't be long before we find it in every sporting goods shop in the country. My choice after extensive testing with the metal match and the magnesium fire starter is clearly the magnesium fire starter. I recommend it highly to all Boy Scouts, backpackers, boaters, campers, fishermen and city survivalists, even to keep in the glove compartment of your car. You never know when an emergency may arise.

Metal matches are commonly available in most camping outlets for under $4. The magnesium fire starter retails for $4.95–$5.95, but it is not yet widely available.

We in the city have taken fire for granted. We need to become adept in the basic survival arts. Fire should not be taken for granted, not even in the urban wilderness.

SURVIVAL CLOTHING

A man who was hopelessly lost heard a helicopter in the distance. He remembered that his undershorts were white. He snatched them off immediately, attached them to a long stick, and actually flagged down the helicopter. At least one article of white or very light clothing, larger than a handkerchief, could be useful in emergency situations.

If a city emergency arises while you are away from home, you probably will not have a chance to go home and change into suitable survival clothing. Those of us who rarely expect the unexpected are precisely the ones caught unprepared. Unexpected situations can and do happen all the time. This test will tell you how well you would do if you were stranded outdoors, dressed in the clothing you're wearing now!

SURVIVAL CLOTHING TEST

1. Are your pants or trousers so tight they restrict your movement?
2. Do your pants have cuffs?
3. Are any of your garments made of wool (either 100% virgin or mixed with no more than 20% other fabrics)?
4. Does your shirt or blouse have pockets?
5. Do your pockets (shirt, blouse or pants) button up?
6. Are any of your clothes blue?
7. Are you carrying a large kerchief (preferably 100% cotton)?
8. Are you wearing flat-soled shoes?
9. Are your shoes made with leather uppers and rubber soles?
10. Do your shoes fit properly with adequate space for toe movement?
11. Are you wearing a leather belt with a sturdy buckle?

Answers:
1. Yes, 0; No, +5. Tight pants inhibit quick and agile movements. If tight clothing tears, it has no protective value. Several layers of loose fitting clothing are better than heavy clothes which cannot adjust to climate changes.
2. Yes, 0; No, +5. Cuffs can get caught and cause hazardous falls.
3. Yes, +10; No, 0. Unlike cotton, wool provides insulation when it is wet. Wool socks are absorbent but insulating, nonflammable and water resistant.
4. Yes, +2; No, 0. Pockets are essential to carry versatile survival items (a dime, knife, fire starter, keys, etc.).
5. Yes, +2; No, 0. Buttons on the pockets help secure valuables.
6. Yes, 0; No, +2. Blue clothing attracts mosquitoes, especially when damp with sweat.
7. Yes, +5; No, 0. Handkerchiefs can be used for many makeshift items, such as towels, bandages or filters. The tougher, more durable cotton kerchief is more advantageous in survival situations.
8. Yes, +6; No, 0. Flat shoes are more conducive to running, jumping, climbing and distance walking than the crippling high heels of popular fashion. Flat shoes arright the spine, while raised heels throw the body forward in a spine-curving posture.
9. Yes, +5; No, 0. Rubber soles are best for gripping. Leather uppers are sturdier and last longer than cotton (sneakers) or vinyl.
10. Yes, +5; No, 0. Proper-fitting shoes will allow you to wear wool socks, without discomforting toe squeezing. Feet tend to swell with extensive walking. Shoe sizes proper for sedentary, smoothly paved city life will cripple under extended periods of walking.
11. Yes, +8; No, 0. A leather belt will make a rope for climbing purposes with the sturdy buckle end secured to a tree. Vinyl and fancy belts are worthless substitutes in emergencies.

Add up your points. Your clothing is survival geared, emergency-ready, if you get 40; not-so-ready with 25; and totally unprepared if you get 10 points. Emergency clothing suitability can mean the difference between surviving and giving up.

PANTS

Pants are an excellent survival tool which can be used immediately, whenever the need arises.

A friend was saved by his pants once, when he was swept out to sea by riptides. He remembered reading in a Boy Scout manual that a wet shirt or pair of pants could be inflated with air and used as a float. He took off his shirt, and with his last bit of strength, whipped it over his shoulder into the water in front of him. The wet shirt captured some air. Holding the shirt openings closed, he was able to stay afloat with his slightly inflated shirt. Once he caught his breath and rested some, he began to swim slowly back to shore to safety.

After I heard of his experience, I was eager to try this method myself. I went with friends to a private pool and we tried this technique. I practiced taking off my pants under water, capturing air in both my pants and shirt. I found that the pants held air far better than the shirt, because shirts have so many air holes.

1. Don't be afraid to go under water; otherwise getting your pants off will take too long.
2. Once the pants are off, tie simple overhand knots at the bottom of each leg.
3. Zip up the zipper (no easy task when in water and trying to stay afloat at the same time).
4. Pull the pants out of the water and throw them from behind, over the shoulder, and down into the water in front, capturing enough air to fill the cavity of the pants.
5. Close the waist band with one hand to retain air.
6. Hold the pants under your right arm, letting your body relax and float with ease.
7. Gradually move the pants so that each "air wing" (pant leg) is on either side of the head as you lie on your back.

WOOL

Wool is a remarkable fiber. History records the use of wool as far back as 300 B.C. Herding sheep flocks was one of the earliest stages of cultural development, between barbarism and civilization.

Wool's value as a survival fabric is unmatched, except for down, because of the chemical structure of the fiber. Abrasion-resistance and water repellency make wool a survival tool. Wool can be a lifesaver, keeping you warm even if you are wet. Have you ever been caught in a cold rain in a pair of thin polyester pants? I have, and the experience is cold, clammy and uncomfortable.

The membrane protecting the scales is nonprotein so that water is not attracted to the fiber's surface. Wool can absorb a great deal of moisture, but it absorbs slowly: it takes a long time for wool to feel damp.

Air spaces increase the protective and insulating power of the wool fiber. Wool is 80% air; *air space, not cloth, insulates and keeps you warm!!*

Outside scales have a protective membrane. Although the scales cause skin irritation for some people, the scale covering gives wool abrasion-resistance and water repellency. If you have experienced skin irritation with wool, try lining the main part of the piece of clothing with silk. This will keep the fabric absorbent and add to its overall strength.

Wool is the most elastic major fiber. It stretches 25% to 35% of its length before breaking. Elasticity and natural "crimp" make wool a giant molecular coil spring. When a wool fiber is stretched, the crimp comes out; when released, the crimp returns. This quality is very important in wrinkle resistance and insulation.

Wool is lightweight, warm, easy to shape and keep pressed, naturally water repellent, and a lifesaver if you ever need it.

THE KERCHIEF

Use the common handkerchief for much more than nose blowing. The kerchief is a far greater survival tool than most citydwellers realize. I know you will find many uses as you begin to carry and use one.

The hobo wore a large kerchief around the neck and drew it over the nose and mouth to protect against smoke or train engine soot, or possibly to conceal his identity. Cowboys of the West used the kerchief to prevent "biting the dust" while riding horses.

The kerchief will make a suitable washcloth, potholder, coffee filter, hat or scarf in both emergency and camping situations. Hikers often use the kerchief to protect the head from the sun on a hot day. Tie one around your neck on cold days for extra warmth. The kerchief will absorb sweat from the forehead. Wrapped around the feet, the kerchief will provide protection and insulation.

Two kerchiefs sewn together will provide a makeshift bathing suit top. Tie all four corners together and make a bag or pouch. Carry several kerchiefs for various uses.

The kerchief is used when necessary as a water filter to remove dirt, leaves or dead bees from water. When you are lost, a bright kerchief will serve as a passable signalling device to flag down help. Tie several kerchiefs together and make an emergency belt, if you should need one. In more unusual cases, the kerchief can be torn into strips and used to secure posts for an emergency lean-to shelter.

Several first aid techniques require cloth or binding material. External bleeding requires direct pressure to the wound, preferably applied with a clean bandage or cloth, such as a kerchief, between your hand and the wound. A kerchief comes in handy as an ace bandage for tying or securing a splint, or as a dressing for cuts and wounds. The kerchief makes a constricting bandage to slow the lymph flow for snake bites. (*Never make a tourniquet unless the limb must be sacrificed.*)

You need not pay outrageous prices for a mere square piece of cotton fabric. Any piece of cotton clothing or material cut 2½" square will suffice to make a kerchief.

SHOES

One of the biggest booms in the shoe industry in recent years has been with the Kelso Negative Heel Earth shoe. Designed in 1957, the shoe was based on the natural footprint in soft sand: the heel is lower than the ball of the foot. Walking barefoot on concrete or wearing heeled shoes, the spine is forced into an unnatural position.

The Earth shoe was designed with both a longitudinal arch and a metatarsel arch for the ball of the foot. This sturdy leather shoe has a durable Vibramlike sole and a patented negative heel with arches.

When I first purchased a pair of Earth Shoes, several imitations were already cropping up. The closest imitation lacked a metatarsel arch. After glorious success in the negative shoe industry, the Earth Shoe company went out of business, and we are left with imitations. I have not tried any negative heel shoes other than Kelso Earth Shoes, but I suspect that some of the imitations may be as good.

SUNGLASSES

Protecting our eyes is another important aspect of survival clothing. Sunglasses are necessary year round, since the glare from snow and ice in the winter can be harsher than the reflection from sand and water in the summer. Don't wait for your eyes to hurt to begin wearing your sunglasses. Carry sunglasses as protection against the low afternoon sun, glare on cars, and as protection from flying dust and dirt. Sunglasses are useful in the winter snow, in the desert, and whenever bright and sunny conditions prevail.

Glare makes the pupil of the eye contract. As the glare's intensity increases, the eyes automatically squint in an attempt to block out ultraviolet and infrared rays; lines and wrinkles appear around the eyes as they move and work constantly. If not adequately protected, our eyes can use up a surprising amount of energy, resulting in, at best, mild exhaustion.

The eyes are particularly vulnerable to the brilliance of the mountain skies, where they can be painfully burned or damaged permanently, if unprotected. Snowblindness can occur even on cloudy days, when the unprotected eye is exposed to glare from the snow. Symptoms of snowblindness are redness, burning, watering or sandy feeling in the eyes, headaches, poor vision, and seeing halos when looking at lights. Symptoms may not appear until four to six hours after exposure. Snowblindness should be treated by staying in a dark shelter or wearing a lightproof bandage over the eyes. The pain can be relieved by putting cold compresses over the eyes. Most cases recover within about 18 hours with no medical treatment.

An emergency substitute for sunglasses is a piece of wood, leather or other material with narrow eye slits cut out. An eyeshade works particularly well in a blizzard; narrow slits can be brushed off, whereas glasses become frosted over.

Shop carefully for sunglasses. Up to 60% of the sunglasses on the market have optical defects which will tire or even harm the eyes. Most dark glasses will filter the infrared and ultraviolet rays adequately; dark green and neutral gray glasses are often recommended by eye specialists. To test the quality of sunglasses, turn them over to catch the reflection of an overhead flourescent light on the inside of the lens. Move the glasses slightly so the reflection travels across the lens; if the light image is wavy and distorted, the lens is faulty.

TIME AND DIRECTION

"What reason is there for me to know how to read a compass or the stars? After all, I have street signs, gas stations, and modern maps to tell me directions. Why in the world would I want to learn a tenderfoot Boy Scout skill?"

No one can deny that when the gas station is open, maps and signs available, and people to give directions, there seems to be little reason to know how to navigate by natural signs. Our cement, asphalt and semblance of order is nothing but a facade that appears to separate us from aspects of animality. We may find ourselves in a situation where we are without the aid of civilization to help us find our way.

Left to our own devices, do we have any survival skills whatsoever? The skills of wilderness survival enable us to be prepared for disasters and emergencies. Even in the city these skills are useful. Consider driving in a strange part of town at night and losing your sense of direction. In such a situation, stars may be helpful, if they are visible; a small compass could help orient yourself when north seems south and west seems east. We can determine time and direction even if we lose our watch and compass. All of nature is a giant timepiece; we need only observe.

INDIAN METHODS

The North American Indians were far from primitive; they possessed advanced ecology in dealing with the land and living *with* it. We can all learn from the wisdom of the Indians. They had no watches or compasses, but they had a keen sense of time and direction, from closely observing and communicating with the natural world surrounding them.

During the day, the sun was their clock. They saw the sun rise in the east, and they watched it set below the western horizon each evening. They faced their tipi openings east to catch the first rays of morning sun. Indians observed certain animals also building their homes with openings facing east, and the pileated woodpecker digging its holes on the east side of trees. They were able to utilize their observations from nature in determining direction.

To get a sense of bearings, the Indian sometimes drove a short vertical stake into the ground, marking the end of its shadow with a rock; and after a short time, marking the end of the new shadow with another rock. A straight line between both rocks is a fairly accurate east-west line. The corresponding perpendicular line is the north-south line. This method works best during midday when the sun is almost directly overhead.

An Indian method for direct travelling to some location was to line up two or three objects in a straight line along the path. Upon reaching each landmark, the Indian would glance back to see how the return trip would look because wilderness trails tend to look entirely different coming and going. They would choose another landmark and travel on, carefully marking each turn with a small pile of rocks, some grass clumps with the heads twisted together, or even broken or scarred branches.

Indian senses came into sharp focus at night. Eyes saw the moon, stars and silhouettes of the mountains and distant horizon. Ears listened intently for the sound of a distant stream or camp. Noses smelled keenly the aromas that meant "danger," "this way home" or "water." The Indian in balance and harmony with Self and the Earth relied upon feelings and intuition as guides. They observed long ago that the stars have fixed relationships to each other and "travel" in regular patterns. Further study of these heavenly bodies made it apparent to these early astronomers that the stars could aid greatly in nighttime travel.

Indian navigational methods may have been less than exact, but then "getting there" was only half the experience; they placed as much value on the journey itself. The Indian never said, "I am lost," but rather, "The camp is lost." Their timekeeping techniques did not have the pinpoint accuracy of modern methods, but then the Indians were concerned more with quality of day than quantity. Their feeling about time is summed up in the Indian saying: *"Yesterday is wood, tomorrow is ashes, only today does the fire burn brightly."*

HAND-RECKONING TECHNIQUE

Indians sometimes reckoned "time of day" by how long the sun had been in the sky or how long before it set. They utilized a hand-reckoning technique for telling time which can still be used today. Facing west, (assuming it is afternoon), outstretch one hand, palm facing the sun, fingers horizontal, with the thumb tucked in. By aligning one hand with the sun, and bringing one hand below the other leapfrog fashion to the horizon, you can determine how many hands distance it is until sunset (hours until sunset). One finger would equal approximately fifteen minutes. This hand-reckoning technique was also used in the morning, while facing east, to figure out how long the sun had been in the sky.

For the Indian people whose days were divided by times of light and times of dark, this hand-reckoning technique seems far more practical than clocks. If you have an almanac listing the times of sunrise and sunset, you will be able to determine "city time" more accurately.

NATURAL SIGNALS

The tips of pine and hemlock trees often point east, while the tips of willows, poplars and alders naturally point south. The prevailing winds also affect the way tree tips point, and have to be considered.

Vegetation is larger and more lush on northern slopes than the smaller, more dense vegetation of the southern slopes. Flowers commonly face east and south. In the northern hemisphere, since the sun appears to move across the sky from east to west and slightly to the south, all the flowers face a more-or-less southerly direction. Plants yearn for the light of the sun. The Indians referred to several plants as compass plants, since almost all their leaves point east and west or north and south.

Spiders build their webs facing south, since the south is usually warmest and dryest. You will find ant hills on the southern warm side of trees and other objects. Indians determined the north side of a hill by the lack of noise and the presence of moisture and moss, from longer snows, water retention and less sun. The exposed south side, on the other hand, is noisy with dried leaves and crackling twigs. Moss always grows on the north side of pine trees, except in a dense forest, where, because of constant and intense moisture, it can grow on all sides.

COMPASSES

Your watch can double as a compass, assuming that it runs accurately. Hold the watch flat and point the hour hand at the sun. The halfway point between the hour hand and the 12 will point south. If your watch is on daylight savings time, the halfway point between the hour hand and the 1 will point south.

If the sun is obscured by clouds, hold a match or twig vertically against the edge of the watch and turn the watch until the match's shadow falls directly on the hour hand. The hour hand will now point to the sun and you can determine south. You should be able to discern at least a faint shadow. Obviously, a watch-compass cannot do all that a more elaborate compass can do. But when you are lost, even a compass cannot tell you where you are; it only helps replace some of the maddening confusion with a sense of order.

For orienting yourself, even in the city, a good compass is a wise investment. One with a luminous dial is best because it can be used night or day. A rugged compass which pins to your clothing is desirable, because it is readily available when needed and will not be lost easily. I carry the small, finely-crafted Silva "Huntsman" compass, which measures only 1½" x 2" when closed. This excellent compass is designed ingeniously to double as a sun dial.

I recommend carrying a second compass to check the accuracy of the first. When it is obvious that one is not working properly, you should follow the one whose needle quivers most freely. Make sure you are not close to metal objects that may throw off your reading, such as stoves, guns and even keys.

The compass needle points magnetic north, not true north. Consult a map of your particular locale to find the difference in degrees between magnetic and true north. East is 90° from north, figuring clockwise on the compass. South is 180° and west is 270°.

SUN CLOCKS

Here's how to construct a crude, but relatively effective, sundial. First, drive a stake *(gnomon)* into the ground at night, making sure that it points directly to the North Star. This is done most easily by sighting the North Star over the tops of two vertical stakes. Align the angle of the gnomon by positioning it over the top of the two stakes. In

212 CITY SURVIVAL

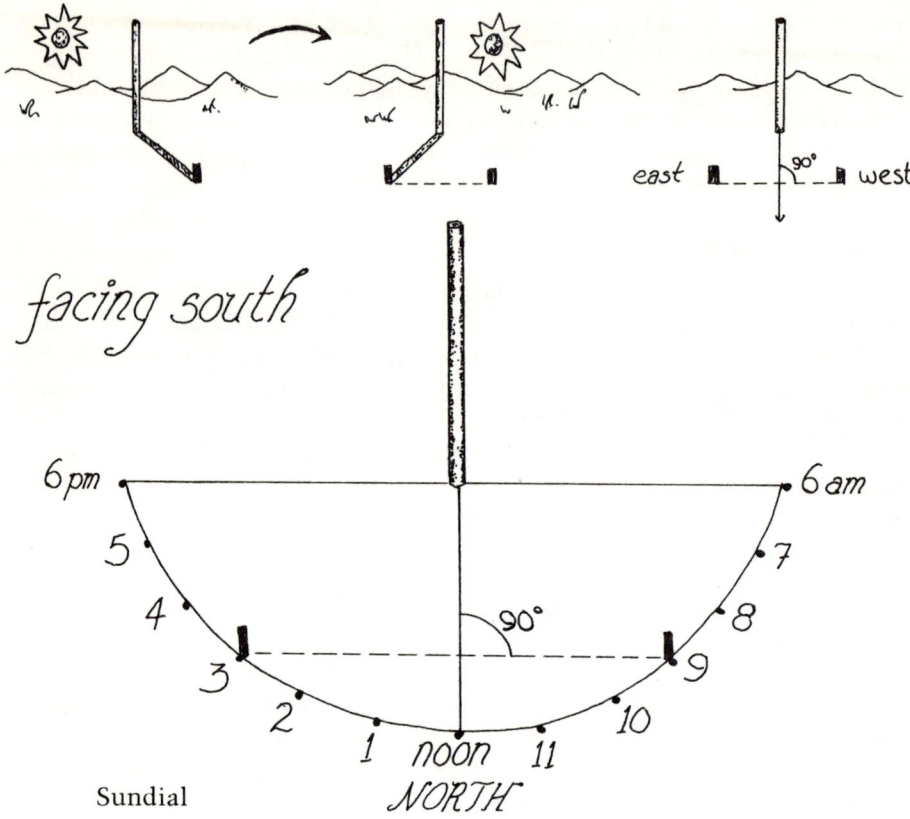

Sundial

the morning tie a string to the base of the gnomon. The string should be the length of the gnomon's shadow. Tie a marker stick to the end of the string. Keeping the string tight, scratch a semicircle around the north side of the gnomon with the marker stick. Drive a small stake in the ground at the end of the gnomon's shadow.

Watch the shadow shorten at noon and then grow again. In the afternoon when the shadow touches the semicircle again, drive a stake into the ground at that point. Now draw a line from both stakes on the semicircle. This is your east-west line. At the halfway point between both stakes, draw a line in the gnomon. This is your north-south line. When the shadow lies on the north-south line, it's noon on your sun clock (standard time). When the sun rises, the westernmost shadow on your clock is approximately 6 a.m. The easternmost shadow on your clock, as the sun sets, reads approximately 6 p.m. You can work out the other hours on your clock to fair approximations.

Variations in the sun clock are due, in part, to the change in relationship between the sun and earth throughout the year. June 21st, for example, is the longest day in the year, with approximatly 14½ hours of daylight. On this day the sun will appear to rise and set the farthest north. December 21st is the shortest day of the year, with about 9½ hours of daylight, and the sun rises and sets the furthest south. These are important considerations, when trying to achieve greater accuracy with sun clocks.

Sun clocks are not as accurate as electric clocks, but they can be made easily, and are especially fascinating for children. When camping in a group, sun clocks can be made large enough for everyone to use. With practice, you will be able to determine the time by observing the sun's position in the sky.

SUGGESTED READING

U.S. Department of the Air Force. *Search and Rescue Survival.* Superintendent of Documents, U.S. Government Printing Office, Washington, D.C. 20402, 1969. Refer to this pamphlet as AFM-64-5 when ordering it. This book is strong in its plant and navigation sections.

SIGHTS, SMELLS AND SOUNDS OF NIGHT

I once led some night nature hikes which proved to be great learning experiences. We wanted to learn how to gather wild food at night, expand our senses and become more perceptive of our surroundings. In a small pocket of "wilderness" in the city, we began our evening hike.

We saw the stars and identified the constellations of the Big and Little Dipper, Cassiopeia, Orion and others which would help us with night navigation. In this world of darkness our eyes saw only shades of gray and black.

We listened to the sweet sound of unfamiliar birds and watched their silhouettes as they flew overhead. An expert birdwatcher in the group, Larry Shaffer, told us how to identify birds by their silhouette and their song. Everyone squinted to see, and listened carefully, trying to differentiate one particular bird sound from among all the sounds of night.

We attempted also to idenfity trees by silhouette and listened to the different sounds made by different trees. We looked closely at plants to observe them in the world of night. We noticed that on some plants the lower leaf surface was whiter than the top (a good point in identification); other plants were identified easily by leaf shape.

Even though we know that after being in darkness for about 30 minutes, human eyes become adjusted like the owl's, we wanted to decrease our overreliance on the eyes during this special night. With our hands we felt the texture of the leaves, the stems and the flowers. By running our fingers over their surfaces, we noticed the differences between tree trunks and rocks.

TRUST YOUR NOSE

Sense of smell played an important part in our glimpse of a different aspect of nature. We smelled all the plants we encountered and quickly learned the nose's vital role in plant identification. Unique were the aromas of sage, California bay, jimsonweed, mugwort, epazote and others. Smells varied as we walked through trees, through grass and into swampy areas.

The nose is a complex instrument which we should always utilize to its fullest capabilities. Trust your nose. It may be telling you that certain animals are present or have recently passed your way. Have you ever noticed the smell of the weather, such as before a rain, before a break in the rain, after a snow, during a wind, or on the early morning of a clear day? The really attentive observer can probably also remember "smelling" fear, danger and dishonesty.

The nose is an integral factor in plant identification. Plants have unique aromas. *Jimsonweed* has the smell of rancid peanut butter. This smell, once registered in the memory, will always help identify the plant. Few have trouble identifying the *California bay* tree, when they smell a few crushed leaves. The *laurel sumac* bush, commonly found in chaparral, has an aroma that works as mosquito repellent, when rubbed over exposed parts of the body. *White sage, black sage, California Sagebrush* and *mugwort*, while similar in smell, are uniquely distinct from one another. *Horehound*, a wooly member of the mint family, preferring dry disturbed areas, is found all over the United States. The appearance of horehound betrays its relation to the mint family; it lacks the characteristic sweet minty aroma. This herb boils up into a powerfully bitter, and incredibly effective, herb tea for coughs and sore throats.

It is very difficult to describe a smell. The easiest thing to do is to compare it to a familiar one. An alert nose is an excellent survival tool. It can detect a potentially dangerous gas leak or smell the dawn, when plants begin manufacturing oxygen again, and become your alarm clock. Wilderness travellers can use the nose to sniff out water or determine if the day's weather will be conducive to travel. But most of all, when you "smell" danger, you can take immediate action to avoid an impending threat.

CITY STARS

You can determine direction even if you can't identify any stars. Pick out a bright star and line it up on the top of two sighting sticks which you have placed vertically in the ground. Observe the chosen star for a few minutes. If it is rising, it is to your east. A star that appears to loop flatly toward your right means you are facing approximately south. A star appearing to swing flatly to the left means you are facing north. You are generally facing west, if your star appears to be dropping.

All the stars appear to rotate counter-clockwise around Polaris, the North Star, and the brightest star in the sky; but, of course, the stars do not. The earth rotates on its axis, while rotating around the

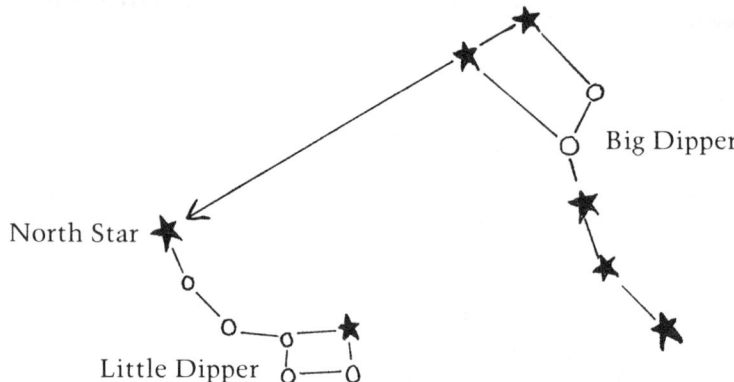

sun, which, along with all the other members of this galaxy, are travelling through space. Think on that one, space partner.

The *North Star* has the reputation of saving more lives than any other star. It is within one degree of true north and can be found easily. To find it, first locate the Big Dipper.

The *Big Dipper* has four stars which form a bowl and three which form a handle. The two stars of the bowl farthest from the handle are called "pointer" stars, because they lie in a straight line with the North Star at approximately five times the distance between the two pointer stars.

You can verify the North Star by finding the *Little Dipper*, since the North Star forms the first handle star of the Little Dipper. The Little Dipper's shape is nearly identical to the Big Dipper, with the handle and bowl of the Little Dipper curving towards the Big Dipper.

Even if you are completely lost, you can find your latitude (your position on the earth north of the equator) with the help of the North Star. Pound two sticks vertically into the ground in a sighting fashion. Line up the top of both sticks so the North Star is sighted over the top of both sticks. Designate a line from the top of both sticks to the ground with a string or stick. The angle of this line to the ground will be your latitude. This method works because at the Earth's equator the North Star is seen on the horizon (0°) and at the North Pole it is seen directly overhead (90°).

Because all stars in the Northern Hemisphere appear to rotate counter-clockwise around the North Star, the Big Dipper is sometimes below the horizon. You can locate the North Star by finding the constellation *Cassiopeia*. Cassiopeia (or the Seated Lady) is the same distance from the North Star as the Big Dipper, but is on the opposite side of the North Star. It looks like a large *M* or *W* depending on its position. Another good constellation to know is *Orion*, named after a mighty hunter. Orion is a bright winter constellation which travels

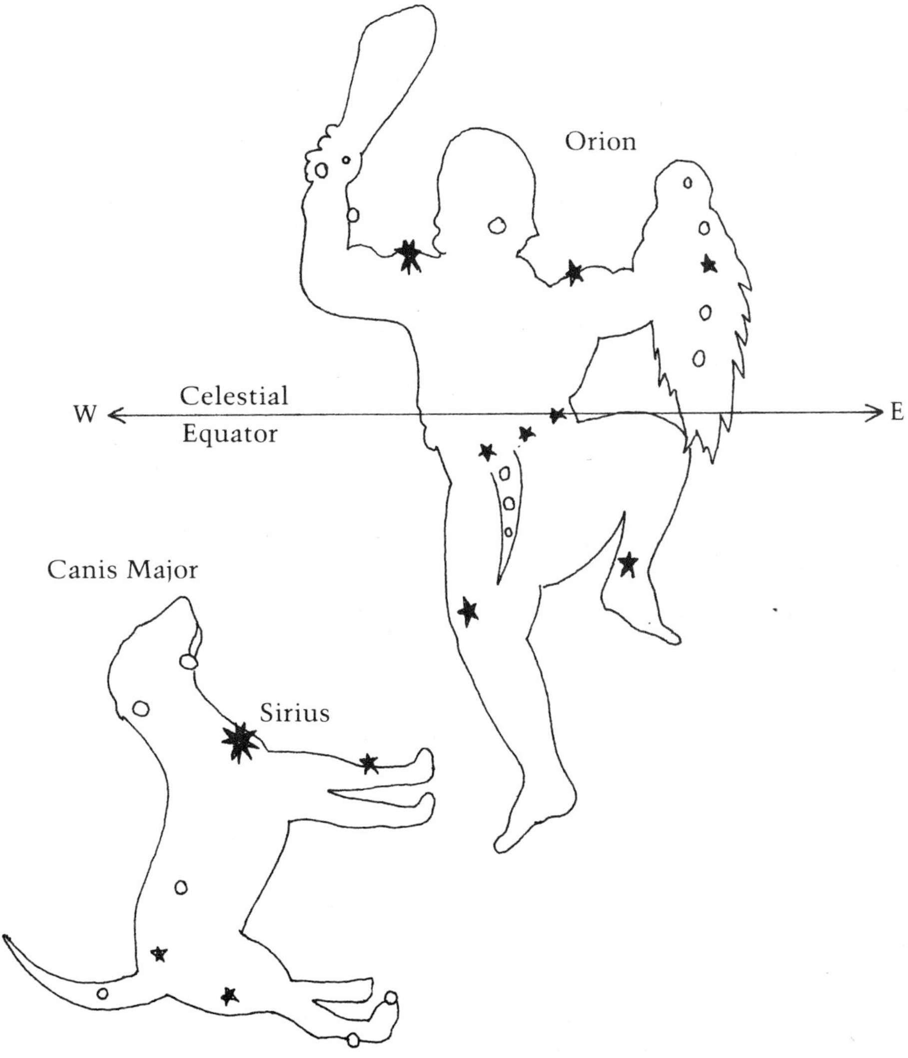

Sights, Smells and Sounds of Night

across the sky in an imaginary line called the *Celestial Equator*, along which both the sun and moon travel. Four bright stars make a lopsided rectangle which encloses three bright stars in a line, making Orion's belt. Below Orion's belt hangs his sword, a line of stars which radiate at an approximate 45° angle to the belt, pointing southeast as you view it. The four stars forming the rectangle mark Orion's two shoulders, his upraised foot and his right knee.

As you face south, while looking at Orion, the star in the belt closest to the right side of the rectangle lies directly on the celestial equator. This means that this star will always rise due east and set due west. The two stars that make up the right half of the lopsided rectangle point north.

Orion is a rich source of stories. The great hunter Orion, son of Neptune, was a favorite of the huntress Diana. But Diana's brother, Apollo, became Orion's enemy and arranged Orion's death. Sorrowful Diana put Orion in the sky, where he remains a giant among constellations. Orion holds a club in his right hand, ready to defend himself against the apparent attack of *Taurus* the bull, another nearby constellation. Orion holds up a lion skin in his left hand as a shield, and his dog (the constellation *Canis Major*) stands below him, ready to serve and protect. Canis Major is easy to locate, since the star *Sirius*, at the point of the dog's neck, is the brightest star in the sky. Tonight be sure to go outdoors and view this spectacle.

SUGGESTED READING

Baker, David. *The Larousse Guide to Astronomy.* New York: Larousse and Company, Inc., 1978. Color-illustrated.

Kyselka, Will; Lanterman, Ray. *North Star to Southern Cross.* Honolulu: The University Press of Hawaii, 1976.

Moore, Patrick. *The A-Z of Astronomy.* New York: Charles Scribner's Sons, 1976.

WEATHER FORECASTING

When you're unable to hear TV or radio weather forecasts, if such forecasts are unavailable by newspaper or telephone, or just for the joy of figuring it out yourself, you can read weather signs from nature.

NATURAL INDICATORS

Tree leaves often show their bottoms before a rain. Leaves grow according to the prevailing wind; and because storm winds generally come from a different direction than the prevailing wind, the leaves turn over. Smoke from a fireplace or campfire curls down with the lowering pressure before a rain: rising smoke generally foretells fair weather.

When the air is dry and the sky is clear, the air becomes heavy with coldness, because there are no clouds to reflect the heat from the earth's surface. Coldness, just like water, will settle in the lowest places, covering everything with dew. Dew on the grass at night or early morning will almost always indicate fair weather.

> When the dew is on the grass,
> Rain will never come to pass;
> When the grass is dry at night,
> Look for rain before the light.

Birds perch more and fly lower before a storm, because the low-pressure atmosphere makes flying more difficult. Birds and fowl fly higher in the sky in good weather.

In the northern hemisphere the winds circling low counterclockwise generally bring inclement weather. If you face the wind, the low-pressure center will be on your right, meaning that the storm will be to your right. This rule is not applicable with sea breezes.

Bark on old trees often tightens in cold and dry air, sometimes causing literally thousands of snapping sounds. Fading stars may sometimes indicate a storm from the thin veil of clouds moving in.

When distant sounds seem loud and hollow, rain may be on the way. Sound waves bounce off the lowering cloud ceiling. Even odors seem stronger in the low-pressure atmosphere that precedes a rain.

"Red sky in morning, sailors take warning; red sky at night, sailor's delight." Most of you would be surprised to learn that Jesus said that about two thousand years ago.

> When it is evening, it will be fair weather: for the sky is red. And in the morning, it will be foul weather today: for the sky is red and lowering. Oh ye hypocrites, ye can discern the face of the sky: but can ye not discern the signs of the times? *Matthew* 16:2-3

Rings around the moon are called halos, which we see from a refraction of light on bodies of moisture in the upper atmosphere. By indicating the presence of moisture, the halo tells us that the moisture may be precipitated as rain or snow. If the pressure is falling, a halo indicates rain within 48 hours and is right about 75% of the time. The location and direction the clouds are moving and how the pressure is changing will need to be considered.

Crickets can be used as thermometers. Count the number of chirps in 15 seconds and then add 40 to get the degrees Fahrenheit. For better accuracy you may want to divide the number of chirps in a minute by four. Another method, which gets the same results is to count the number of chirps per minute, subtract forty, divide the result by 4, and add 50. This number is also the degrees Fahrenheit.

It is possible to tell the distance of a storm by the time lapse between lightning and thunder. Thunder is an audible compression wave, caused by the rapid heating of air by a return lightning stroke. Downward leader strokes go only a hundred to a thousand miles per second: the stroke you see (because the electricity heats the air white hot in returning to the cloud) goes about 87,000 miles per second (half the 186,282 miles per second of the speed of light).

You see the lightning flash, no matter how far away, .005 to .002 seconds after it heats the air briefly. Sound travels at only about 1,088 feet per second at 32° F sea level. It goes faster in warm air (1,130 miles per second in 68°), but slower at high altitudes, and not at all in a vacuum. We can figure that a storm is roughly a fifth of a mile away for each second that passes between the time we see lightning and hear thunder; if we see lightning and then hear thunder about 5 seconds later, then the storm is a mile away.

CLOUDS

Clouds are most useful for short-term weather prediction. One must observe the manner in which clouds are changing—the temperature, wind changes, time of year and geographical location—to "read" clouds accurately. The four types of clouds are cirrus, cumulus, stratus and nimbus. Clouds overlap and modify as they descend, forming combinations of the four types.

Cirrus clouds are high in the atmosphere and appear featherlike, with slender fingers running out in all directions. They are like thin wisps or curls of hair, and are sometimes elongated by the wind. A vivid blue sky behind morning cirrus clouds usually means the clouds will disappear before noon and leave a clear sky. An abundance of cirrus clouds, however, can mean rain in 24 hours.

Cumulus clouds are huge, white, rounded masses with flat bases, having a soft fluffy appearance. Appearing in a blue sky, cumulus clouds generally mean fair weather; building vertically in large banks, they warn of showers (usually in the evening or late afternoon).

Stratus clouds are generally long, horizontal bands which form near the earth's surface. Massing together, they cover the sky. Although they may bring drizzles, they seldom bring rain. Stratus clouds are often the common coastal fog—layered clouds flat on both top and bottom—which ordinarily dissipate in sun-warmed skies. When a low-pressure area is approaching, stratus clouds are often followed by nimbostratus clouds, denser clouds that bring rain or snow 75% of the time, and within a few hours.

Nimbus clouds are low masses which have no definite contour. They are the clouds from which rain or snow is usually falling.

Cirrostratus clouds usually indicate a storm within a day or so. When the sky behind cirrus clouds is gray, the formations usually develop quickly into the denser cirrostratus. Cirrostratus clouds are made up of ice particles and often resemble white veins. These clouds are commonly responsible for halos of both the sun and moon.

Cirrocumulus clouds seem to be not quite cumulus clouds and not quite cirrus clouds; they are the halfway point between both. They generally appear after storms and almost always indicate fair weather. They are commonly called mackerel or a "mackerel sky" because their pattern resembles the colorations on a fish's back.

Altocumulus is another fair weather cloud. Because these clustered white mounds are often cut into flakes by vertical air currents, they closely resemble the mackerel sky.

As a storm grows closer, cirrostratus clouds thicken and form *altostratus* clouds which allow the sun to shine through as a bright spot or blob. Altostratus clouds appear as a gray or dull blue haze, and are followed by steady, but not hard, snow or rain.

Stratocumulus clouds are usually twisted like a twisted breakfast roll or pastry. They rarely bring rain, and usually dissipate into cumulus clouds.

Cumulonimbus, another variant of the cumulus cloud, is a towering cumulus cloud with cirrus at the top and nimbus at the bottom. They drop rain or hail.

CLOUD FORMATIONS

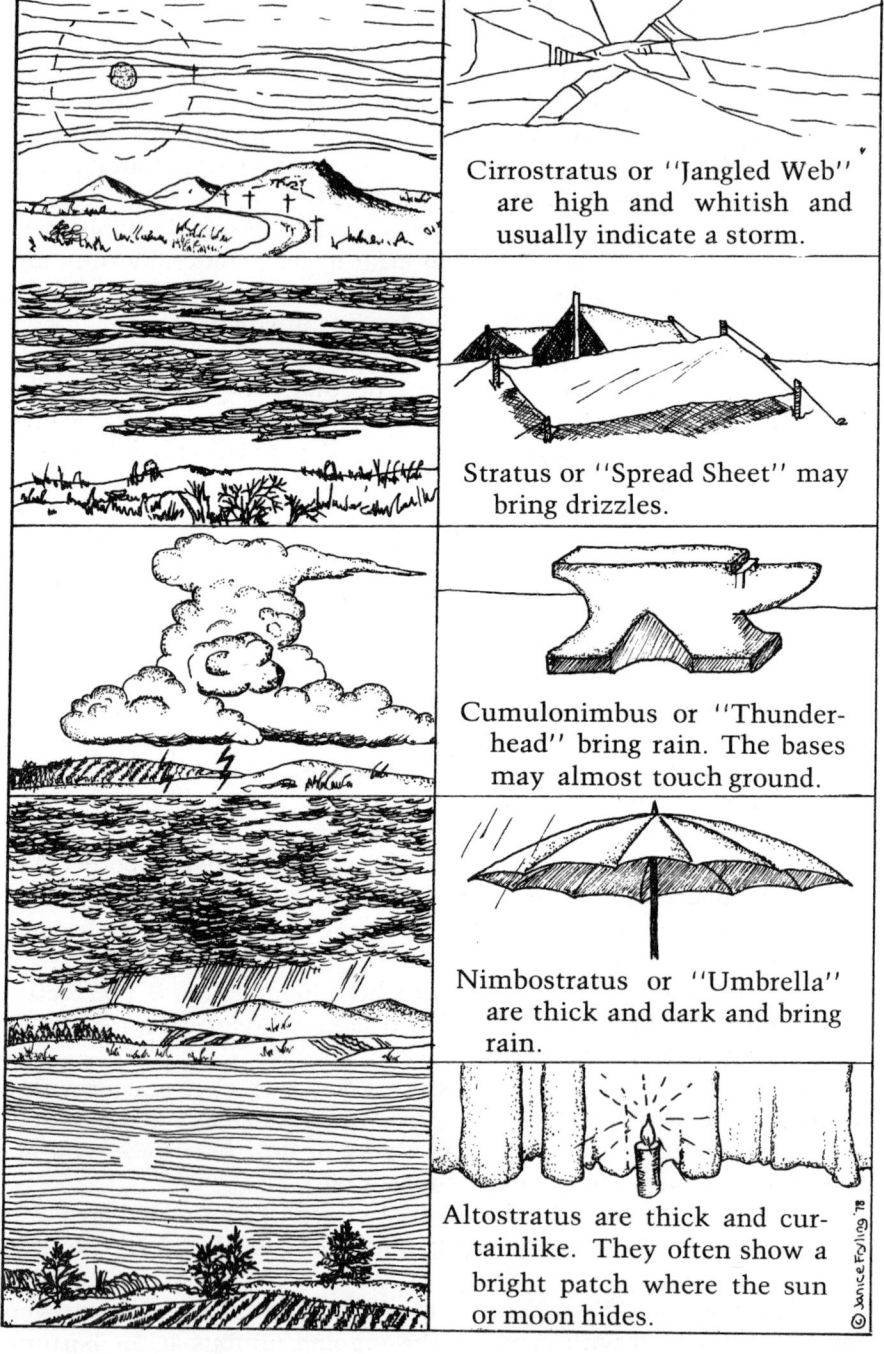

Cirrostratus or "Jangled Web" are high and whitish and usually indicate a storm.

Stratus or "Spread Sheet" may bring drizzles.

Cumulonimbus or "Thunderhead" bring rain. The bases may almost touch ground.

Nimbostratus or "Umbrella" are thick and dark and bring rain.

Altostratus are thick and curtainlike. They often show a bright patch where the sun or moon hides.

Weather Forecasting

Cumulus or "Wool Pack" are large and puffy and generally mean fair weather.

Cirrus or "Feather" are the highest of the formations (5-6 miles high).

Altocumulus or "Sheep" appear in groups or large masses (3-4 miles high).

Stratocumulus or "Twist" rarely bring rain and they usually dissipate into cumulus (1 mile high).

Cirrocumulus or "Mackerel" appear as small flakes arranged in groups or lines (4 miles high).

SUGGESTED READING

Gales, Donald Moore. *Handbook of Wildflowers, Weeds, Wildlife, and Weather of the Palos Verdes Peninsula.* San Pedro, CA: Caligraphics Printing and Publishing, 1974. Granted, the title indicates that this book has limited use, but it can actually be used widely over most of the Western United States, and to a limited extent throughout the United States. The drawings by the author are a pleasure to view—done with many annotations, so that one really "knows" plants after reading this book. Gales received an Achievement Award from the American Meterological Society for the chapter on weather and climate.

Lee, Albert. *Weather Wisdom.* Garden City: Doubleday & Company, Inc., 1977. Illustrated practical book; a compilation of facts and folklore of natural weather prediction. A beautiful book as well as useful.

A GLOSSARY OF PLANT PARTS AND SHAPES

LEAF PARTS

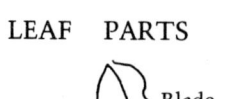

LEAF SHAPES

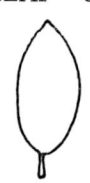
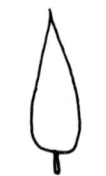

Elliptic   Lanceolate   Linear

LEAF SHAPES (Cont'd)

Oblong   Ovate   Lyrate   Cleft   Orbicular

LEAF MARGINS (SIMPLE)

Entire   Serrate   Lobed   Crenate   Incised   Dentate

COMPOUND MARGINS

Pinnately Compound   Palmately Compound

LEAF ARRANGEMENTS

Opposite   Alternate   Whorled

LEAF ATTACHMENTS

Petiolate   Sessile   Decurrent   Basal

FLOWER

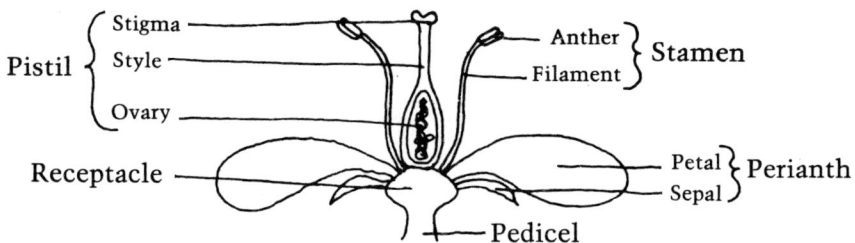

INFLORESCENCES

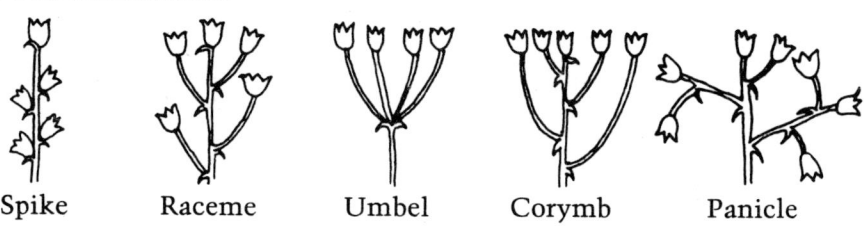

INFLORESCENCES (Cont'd.)   ROOTS

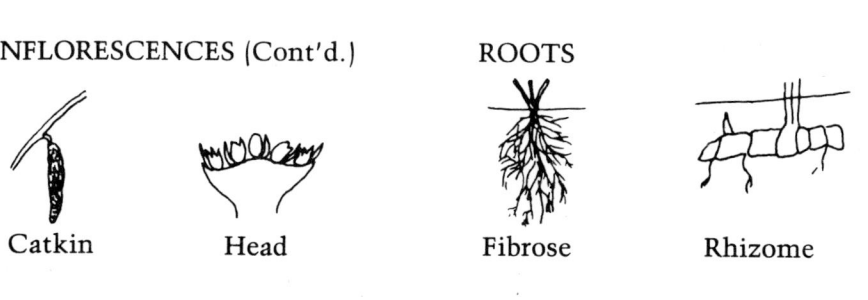

ROOTS (Cont'd.)

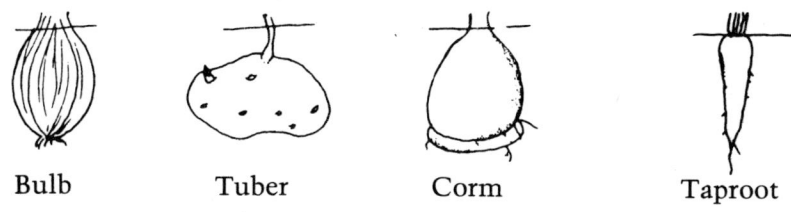

FRUIT

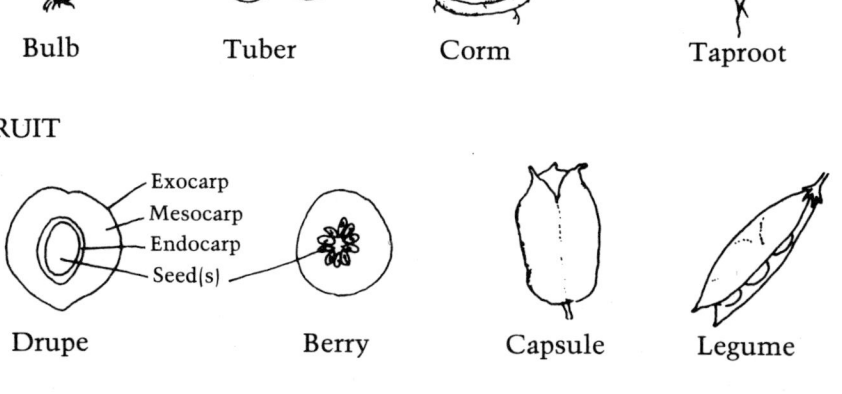

Christopher Nyerges is the author of *A Southern Californian's Guide to Wild Foods* and a syndicated columnist in over eighty newspapers for hiking, biking and wild foods. His articles have appeared in *Organic Gardening, Mother Earth News, Well-being, Home Magazine,* and more. Nyerges is a naturalist who gives lectures to many organizations on various subjects, including recycling and Indian lore. He is most well-known for his valuable information on wild foods. He conducts wild food hikes regularly throughout the Southern California area, to teach others how to identify, gather and use wild edible plants. Christopher Nyerges is a board member of White Tower, Inc., a nonprofit organization that researches, publishes and educates in all aspects of survival.